Sustainable Development:
The Challenge of Transition

Demographers predict that world population will double to around 12 billion people during the first half of the twenty-first century and then begin to level off. Based on this scenario, *Sustainable Development: The Challenge of Transition* examines what societal changes must occur over the next generation to ensure a successful transition to sustainability. A successful transition must provide for the needs of these people within the constraints of the natural environment.

An array of prominent authors presents a broad discussion of the dimensions of sustainable development: not just economic and environmental, but also spiritual and religious, corporate and social, scientific and political. Unlike other books on the subject, this volume provides insightful policy recommendations about how business, government, and individuals must change their current values, priorities, and behavior to meet these challenges. These types of changes ordinarily take many decades to occur, so it is important to begin making them now, before the problems are overwhelming and more costly.

This volume will appeal to scholars and decision makers interested in global change, environmental policy, population growth, and sustainable development, and also to corporate environmental managers.

Sustainable Development: The Challenge of Transition

Edited by

JURGEN SCHMANDT

*University of Texas at Austin and Houston
Advanced Research Center, Texas, U.S.A.*

and

C.H. WARD

Rice University, Houston, Texas, U.S.A.

with the assistance of

MARILU HASTINGS

Houston Advanced Research Center, Texas, U.S.A.

CAMBRIDGE
UNIVERSITY PRESS

PUBLISHED BY THE PRESS SYNDICATE OF THE UNIVERSITY OF CAMBRIDGE
The Pitt Building, Trumpington Street, Cambridge, United Kingdom

CAMBRIDGE UNIVERSITY PRESS
The Edinburgh Building, Cambridge CB2 2RU, UK   http://www.cup.cam.ac.uk
40 West 20th Street, New York, NY 10011–4211, USA   http://www.cup.org
10 Stamford Road, Oakleigh, Melbourne 3166, Australia
Ruiz de Alarcón 13, 28014 Madrid, Spain

© Cambridge University Press 2000

First published 2000

Printed in the United Kingdom at the University Press, Cambridge

Typeface TEFFLexicon 9/13 pt   *System* QuarkXPress®   [SE]

A catalogue record for this book is available from the British Library

Library of Congress Cataloguing in Publication data

Sustainable development : the challenge of transition / edited by
Jurgen Schmandt, C. H. Ward.
    p.   cm.
This volume is built upon the foundation of a conference hosted by
Rice University in March 1997.
Includes index.
ISBN 0 521 65305 3
1. Sustainable development.   2. Environmental policy.
I. Schmandt, Jurgen.   II. Ward, C. H. (Calvin Herbert), 1933–   .
HC79.E5S8684   2000
338.9–dc21   99–31467   CIP

ISBN 0 521 65305 3 hardback

Contents

Contributors

Carla V. Berkedal

is an Episcopal priest and has worked in parish ministry since 1980. She is currently Founding Director of Earth Ministry and the Associate Rector at Emmanuel Episcopal Church on Mercer Island, Washington. As Founding Director of Earth Ministry, a nonprofit, ecumenical, environmental ministry located in Seattle, Carla works with individuals and congregations to help them discover the Christian roots of living lightly, and with gratitude, on the earth. Ms Berkedal has extensive experience as an author, teacher, speaker, consultant, and team member of various boards, committees, and coalitions inside the Church and beyond, including the Presiding Bishop's Environmental Stewardship Team, Sustainable Seattle, and People for Puget Sound. Her voice is especially sought after as one to interpret and apply Christian spirituality and ethics to the care of creation. She has also been honored as a Bloedel Reserve Fellow and as the Seattle Audubon Society's "Environmentalist of the Year." Ms Berkedal's educational accomplishments include a BA in Religion and Philosophy from Bowdoin College, Certificates in Theology and Greek from the University of Strasbourg, and a Master of Divinity from Yale Divinity School.

James B. Blackburn

is an environmental attorney and lecturer at Rice University. His law practice involves litigation on air quality, land contamination, risk assessment, environmental auditing, and zero discharge planning, as well as natural resource issues such as wetlands, flooding, and endangered species. Mr Blackburn has taught Environmental Law and Global Environmental Law in the Environmental Science and Engineering Department, and Environmental Planning and Sustainable Development in the Architecture School at Rice University. He received a BA in History, a JD in Law at the University of Texas at Austin and an MS in Environmental Science from Rice University.

FRANCES CAIRNCROSS

is Management Editor on *The Economist*. She was the magazine's first environment editor. Before that, she held posts as principal economic analyst on the *Guardian* newspaper, and as editor of *The Economist's* Britain section. In 1992, she was editor of the World Bank's "World Development Report." Her book, *Costing the Earth: the challenge for governments, the opportunities for business*, is published by Harvard Business School Press and has been translated into Chinese, Danish, Finnish, French, Italian, Japanese, and Spanish. A further book, *Green Inc.*, was published in Britain by Earthscan in May 1995 and in September by Island Press in the United States. Her new book, *The Death of Distance*, is a study of economic and social effects of the global communications revolution and was published by Harvard Business School Press in fall 1997. Ms Cairncross was educated at Oxford University and at Brown University in Rhode Island.

HERMAN E. DALY

is Senior Research Scholar at the University of Maryland, School of Public Affairs. From 1988 to 1994, he was Senior Economist in the Environment Department of the World Bank. Prior to 1988 he was Alumni Professor of Economics at Louisiana State University, where he taught economics for 20 years. He has served as Ford Foundation Visiting Professor at the University of Ceara (Brazil), as a Research Associate at Yale University, as a Visiting Fellow at the Australian National University, and as a Senior Fulbright Lecturer in Brazil. Dr Daly has served on the boards of directors of numerous environmental organizations, including the International Society for Ecological Economics, the Beijer Ecological Economics Institute of the Swedish Royal Academy of Sciences, the World Watch Institute, and is co-founder and associate editor of the journal, *Ecological Economics*. His interest in economic development, population, resources, and environment has resulted in over a hundred articles in professional journals and anthologies, as well as numerous books. In 1996, Dr Daly received the Honorary Right Livelihood Award (Sweden) and the Heineken Prize for Environmental Science awarded by the Royal Netherlands Academy of Arts and Sciences. He holds a BA from Rice University and a PhD from Vanderbilt University.

MALCOLM GILLIS

is the sixth president of Rice University. His first academic post, as an Assistant Professor of Economics at Duke University, was followed by years at Harvard. He returned to Duke in 1984 as Professor of Economics and Public Policy. From 1986 to July 1991, Dr Gillis served as Dean of the Graduate School and Vice Provost for Academic Affairs. From 1991 until June 1993, he was Dean of the Faculty of Arts and Sciences. In recent years, his research and teaching activities have been focused upon two broad classes of issues in their national and international dimensions: fiscal reform and environmental policy. He was the co-founder and chair of the Board of Trustees of the Center for World Environment and Sustainable Development and was also co-founder of the Duke Center for Tropical Conservation. He has authored or edited ten books. In addition, he has published more than sixty articles in journals and books. Dr Gillis has spent his professional life bringing economic analysis to bear on important issues of public policy in nearly twenty countries, from U.S. and Canada to Ecuador, Colombia, Ghana and Indonesia. He has been a frequent consultant to the State of Alaska, the U.S. Treasury Department, the Canadian Ministry of Finance, the World Bank, and the Governments of Colombia, Ecuador, Bolivia, and Indonesia. Dr Gillis received his BA and MA degrees from the University of Florida and his PhD from the University of Illinois.

ROBERT W. KATES
is an independent scholar in Trenton, Maine and University Professor (Emeritus) at Brown University. He is an Executive Editor of *Environment* magazine, Co-chair of *Overcoming Hunger in the 1990s*, Distinguished Scientist at the George Perkins Marsh Institute at Clark University, Faculty Associate of the College of the Atlantic, and Senior Fellow at The H. John Heinz III Center for Science, Economics and the Environment. His current research and professional interests include the prevalence and persistence of hunger; long-term population dynamics; sustainability of the biosphere, climate impact assessment; and theory of the human environment. Dr Kates is a recipient of the 1991 National Medal of Science, the MacArthur Prize Fellowship (1981–5), the honors award of the Association of American Geographers, and an honorary degree from Clark University. He is a member of the National Academy of Sciences, the American Academy of Arts and Sciences, a foreign member of the Academia Europaea, and a fellow of the American Association for the Advancement of Science. In 1993–4, he was the President of the Association of American Geographers. He holds a PhD degree in geography from the University of Chicago.

BRUCE W. PIASECKI
is a professor at Rensselaer Polytechnic Institute's School of Management and Business Administration and director of their Master of Science degree program in Environmental Management and Policy. Prior to his current position, Dr Piasecki was Associate Director of the Hazardous Waste and Toxic Substances and Management Center at Clarkson University. Internationally, he has been a recipient of various grants, including the German Marshall Fund of the United States for his research throughout

Europe on environmental management developments. In 1981, Dr Piasecki founded the American Hazard Control Group, a business development organization providing legal, planning, policy, and communication advice on environmental management decisions. He remains President of the AHC Group, based in Troy, New York. His clients have included the American Plastics Council, Consolidation Edison, Allergan, Energy Answers Corporation, the New York State Energy Authority, CIBA-GEIGY, the New Jersey Siting Commission, and the Ontario Waste Management Corporation.

JURGEN A. SCHMANDT

is Professor of public affairs at the University of Texas in Austin. He is also Chief Policy Officer and director, Center for Global Studies, at the Houston Advanced Research Center (HARC). His current areas of work include water management, impacts of global change, and policies for sustainable development. His professional career includes work with the German Academic Exchange Service, Organization of Cooperation and Economic Development, Harvard University, and the US Environmental Protection Agency. In Austin, Dr Schmandt participated in the planning and development of a new school of public affairs. He was particularly active in the development of year-long policy research projects which help graduate students gain experience in working with public or private sector agencies. At HARC, he has developed a research center specializing in assessing the regional impacts of global change. He has worked closely with colleagues in Brazil and Mexico on problems of semi-arid regions. Dr Schmandt received his doctor of philosophy degree from the University of Bonn.

STEPHEN H. SCHNEIDER

is a professor in the Department of Biological Sciences, a Senior Fellow at the Institute for International Studies and Professor by Courtesy in the Department of Civil Engineering at Stanford University. He was honored in 1992 with a MacArthur Fellowship for his ability to integrate results of global climate research through public lectures, seminars, classroom teaching, environmental assessment committees, media appearances, Congressional testimony, and research collaboration with colleagues. He also received, in 1991, the American Association for the Advancement of Science/Westinghouse Award for Public Understanding of Science and Technology for furthering public understanding of environmental science and its implications for public policy. Dr Schneider's current global change research interests include climatic change; global warming; food/climate and other environmental/science public policy issues; ecological and economic implications of climatic change; climatic modeling of paleoclimates and of human impacts on climate, e.g., carbon dioxide "greenhouse effect" or environmental consequences of nuclear war. He is also interested in advancing public understanding of science and in improving formal environmental education in primary and secondary schools. Dr Schneider received his PhD in Mechanical Engineering and Plasma Physics from Columbia University in 1971.

JEFFREY R. VINCENT
is a Fellow of the Institute at Harvard Institute for International Development (HIID). He is also a lecturer on economics in the Department of Economics and an adjunct lecturer in public policy at the Kennedy School of Government. He joined HIID in 1990 after teaching for three years at Michigan State University. He is a resource economist with special interests in tropical forest policy, national income accounts and the environment, and environmental management in newly industrializing and transition economies (in particular, pollution policy). He was resident in Malaysia during 1992–93 as leader of that country's first comprehensive research project on environment and development, and he has worked extensively elsewhere in Asia as a researcher, policy adviser, and workshop lecturer. From 1994 to 1998, he was director of HIID's Newly Independent States Environmental Economics and Policy Project, which advised the governments of Russia and the Central Asian Republics on natural resource and environmental policy.

C.H. "HERB" WARD
is the Foyt Family Chair of Engineering in the George R. Brown School of Engineering at Rice University. He is also Professor of Environmental Science and Engineering and Ecology and Evolutionary Biology. Following 22 years as Chair of the Department of Environmental Science and Engineering at Rice University, Dr Ward is now Director of the Energy and Environmental Systems Institute (EESI), a university-wide program designed to mobilize industry, government, and academia to focus on problems related to energy production and environmental protection. He is also Director of the Department of Defense Advanced Applied Technology Demonstration Facility. For the past 14 years, he has directed the activities of the National Center for Ground Water Research. He is also Co-Director of the EPA-sponsored Hazardous Substances Research Center/South & Southwest. Dr Ward is currently Vice President of the U.S. National Committee of the International Water Resources Association and is the co-founder and Editor-in-Chief of the international journal, *Environmental Toxicology and Chemistry*. Dr Ward earned his PhD from Cornell University. He also earned the Master of Public Health in environmental health from The University of Texas School of Public Health.

Foreword

Each human generation chooses not only its place on the planet, but also the state of the planet it will leave in its place. If trends persist, the legacy of the current occupants will be mixed. Some but not all of our descendants will enjoy increased material well-being and opportunities for fulfilling lives, and the world as a whole will be a warmer, more crowded, more consuming, less biologically diverse, and more stressed place. These latter trends can not be supported indefinitely. They will limit the choices of future generations, and they will affect the developed as well as the developing countries. Within the early decades of the next century, there must be a transition to a new path of human development. Shifting course will require that people meet their wants and needs in ways that move away from degrading the planet's life support systems towards sustaining and restoring those systems, and that move away from widening disparities in human welfare towards improved living standards worldwide that reduce or eliminate hunger and poverty. We need to begin to pursue this path today, if we are to leave a more sustainable world for future generations.

How can we make better choices about environment and development to achieve the goals of sustainable development? Consider, for example, the requirements for meeting the goal declared at the World Food Summit of 1996 of reducing the number of undernourished people to half their present level of approximately 800 million no later than 2015. This goal can be achieved only with significant expansion of and efficiency in agricultural production and food distribution in developing countries. Thus, we need to utilize the advances of the modern biological revolution as well as gain a better understanding of the interactions among the natural and social factors affecting agricultural production and food distribution. These include soil fertility loss, climate change, air and water quality changes, increasing pesticide resistance, and the range of policies (e.g., for land-use, transportation, energy, market regulation) that influence land management, economies, incentives, and behaviors.

But knowledge by itself is not enough. We must diffuse this knowledge and the ability to use it around the world to advance the prospects of all people for sustainable development. We need to connect what is known to be wise long-term behavior to the everyday decisions we all make and the actions we take. This requires a life-long, intergenerational process of learning and adapting amidst times of turbulence and surprise. Frequent monitoring will be needed to help seize opportunities, respond to threats, and take corrective steps. Science provides much of the knowledge that we need for action. Thus, it is essential that the scientific and technical communities exert leadership in the production and dissemination of the knowledge required to influence choices about the environment and development and to set us on a course toward sustainability.

In late 1994, with the generous support of Mr George P. Mitchell, Chairman and Chief Executive Officer, Mitchell Energy and Development Corporation, the National Academy of Sciences embarked on a Global Commons Project, the goals of which center on sustainable development. The project is being conducted in cooperation with the Houston Advanced Research Center (HARC) and includes a number of activities of both institutions. HARC seeks to engage the private sector in advancing the application of sustainability concepts in business practices and industrial production. This volume, which is built upon the foundation of a conference hosted by Rice University in March 1997 on 'Sustainable Development: Managing the Transition,' represents a contribution of HARC and Rice University to the overall effort.

The Academy complex – the National Academy of Sciences, the National Academy of Engineering, and the Institute of Medicine – with its broad ability to reach into the scientific, medical, and engineering communities both nationally and internationally, is conducting individual studies of sustainability concerned with fisheries, land use, Middle East water supply, and megacities, as well as a comprehensive assessment of the knowledge requirements for a successful transition to sustainability in the early years of the next century. All of these findings will be presented at a Year 2000 Conference on the worldwide transition to sustainability being sponsored by the InterAcademy Panel on International Issues, an umbrella organization for the world's science academies. Through such activities, the Academy seeks to illuminate the critical issues in development, promote science-based responses, and encourage greater involvement of the scientific and technical communities worldwide in working with governments, the private sector, and international organizations and institutions to shape a sustainable future.

The sustainable fisheries project aims to evaluate whether current marine-capture fisheries are sustainable, determine to what degree marine ecosystems are affected by fishing, and assess whether an ecosystem approach to fishery management can help achieve sustainability. The study of the relationship between population and land use is being conducted in partnership with the Chinese Academy of Sciences and the Indian National Science Academy, using case studies of the Pearl River Delta and the Jitai Basin in China; Kerala and the Haryana Punjab in India; and the Florida Everglades and Chicago in the United States. The range in population and land use policies, social and economic status, and histories of development at these six sites provides a diversity that is important for designing creative practical strategies that address the problems of population pressure on land use. An international committee of experts is carrying out a study of sustainable water supplies for the Middle East, focusing on methods developed in the Middle East and elsewhere for avoiding over-exploitation of water resources, and exploring relationships between water supply enhancement and preservation of environmental quality. The study is considering the scientific and technological aspects of such issues as treating municipal wastewater for irrigation and other purposes, desalination, water harvesting, restoration of degraded water bodies, ground water contamination cleanup, and opportunities offered by improved conservation technologies to enhance water quality and prevent resource degradation. The study evaluating sustainable economic development in the rapidly growing megacities of the world focuses on four major issues for megacities: urban employment opportunities; technologies for affordable housing; sustainable water and sanitation services; and affordable, less-polluting transportation. Initial findings and recommendations were presented at the United Nations Conference on Human Settlements (Habitat II) in Istanbul in June 1996.

Last but not least, an overall vision for exploring a sustainable future is being scoped by our Board on Sustainable Development in a far reaching study to develop a research agenda for a transition to sustainability in the twenty-first century. In the Board's view, a successful sustainability transition would require that the world provide the energy, materials, and knowledge to feed, house, nurture, educate, and employ many more people than are alive today – while preserving the basic life support systems of the planet and reducing hunger and poverty. Such a profound and unprecedented transition has no charted course. Science can help provide direction by identifying the energy, materials, and knowledge requirements for the transition; the crucial indicators of unsustainability; the levers of change to move us towards sustainability; and the measurements needed to report on our progress.

The Board's approach is unique. It focuses on the time period of the next two generations when the most important transitions, especially the leveling off of population, will be taking place, and a time when action can make a real difference for outcomes. It chooses a science base rather than politics for determining the needs and requirements for changing the course of human development. It identifies a set of normative goals for meeting human needs, based on international agreements rather than moral decisions. It discusses social and environmental challenges together rather than as separate elements of the development conundrum. It describes a science and technology agenda that draws upon all branches of science – the natural, social and medical sciences, plus engineering. And, it identifies new technologies and institutional needs, as well as tools and incentives, to facilitate social learning and adaptive management. This report was published in late 1999.

Like the Board's report, the present volume explores ways of thinking about the future. It describes how humanity has responded to the challenges of global change with prospects for sustainable development. The contributors represent broad disciplinary expertise and have diverse views of what the future holds. Their efforts reinforce an important lesson. Only with a greatly increased sharing of ideas and solutions – among all sciences, all sectors, all institutions, and all people – will we be able to strengthen our ability to leave behind a sustainable world full of robust possibilities for the many generations to come.

Bruce Alberts
President
National Academy of Sciences

Preface

One of the most important topics for the twenty-first Century is *sustainable development* – development that meets the needs of the present without foreclosing the needs and options of future generations. Sustainable development is a comprehensive concept involving many disciplines, giving it both a philosophical base and a pragmatic approach. It appeals to a wide range of interests from nations to corporations to stakeholders because it combines economic development and environmental protection into a single system.

As the global community enters into the next phase of the evolution of sustainable development, making the transition from theory to action will be the next great challenge. World population is expected to double within the next 50 years, placing enormous demands on resources for food, shelter, and energy. Innovative solutions, incorporating the concepts of sustainable development, must be developed and implemented to mitigate these impending economic and environmental uncertainties and to guide society through the resulting period of transition. Examining this shift from theory to action and developing a plan of action was the framework of the De Lange ♦ Woodlands Conference, which was held March 3–5, 1997 in the Alice Pratt Brown Hall at Rice University in Houston, Texas.

The conference was organized by the Energy and Environmental Systems Institute at Rice University and the Center for Global Studies at the Houston Advanced Research Center in The Woodlands, in partnership with the National Academy of Sciences and the James A. Baker III Institute for Public Policy at Rice University. A distinguished group of world leaders were invited to participate in six conference sessions over a three day period: (1) Sustainable Development: Defining the Twenty-first

Century Challenges; (2) Achieving Ethical and Equitable Leadership; (3) Scientific Issues and Uncertainty in Decision Making; (4) Market Tools: Trade, Pricing, and Signals; (5) Stakeholders, Empowerment, and Dispute Resolution; and (6) Charting the Roadmap: Institutions, Leadership, and Policies. The conference, attended by over 500 people, provided an international forum for vigorous debate of sustainable development issues and timely analysis of the state-of-progress of sustainable development thought and action. Conference papers were published in a proceedings (*Sustainable Development: Managing the Transition*, De Lange ♦ Woodlands Conference, Rice University, Houston TX, 639 pp, 1997).

The De Lange endowment at Rice University was established by Mr and Mrs C.M. Hudspeth in 1991, in memory of her parents, Albert and Demaris De Lange, for the purpose of organizing conferences at Rice featuring top ranked experts and major world figures to address topics of great concern to society. Past De Lange conferences have focused on Julian Huxley, 1887–1975 (1987), Human Impact on the Environment (1991), and Biotechnology: Science, Engineering, and Ethical Challenges for the 21st Century (1994). The Woodlands Conference Series was founded by oilman and developer George P. Mitchell in 1974. Five Woodlands conferences have been held, all dealing with various aspects of global change and sustainable development. At each conference, the George and Cynthia Mitchell Prize for Sustainable Development was awarded. The 1997 prize winners were Marcelo de Andrade from Brazil and nine young scholars from around the world.

This volume is not a proceedings, but it is a product of the De Lange ♦ Woodlands Conference. A distinguished speaker from each of the conference sessions was commissioned to write a chapter highlighting important features of each session but stressing his or her own insights, experience, and conclusions. We hope the product of their labors and our efforts to bring this volume together will contribute to a continuing international dialog on sustainable development.

C.H. Ward
Jürgen Schmandt

JURGEN SCHMANDT *and*

C. H. WARD

1
_____

Challenge and response

Albert Toynbee, in his monumental study of world history, used the concepts of "Challenge and Response" to explain how civilizations rise and fall. He felt that traditional explanations – environment, race, leadership, possession of land, access to natural resources – were wrong or too narrow. Instead, he looked for the underlying cause that explained societal success or failure. By "challenge" Toynbee meant some unpredictable factor or event that posed a threat to the ways in which a group of people had made their livelihood in the past. But "challenge" was not all negative. It carried in it the germ of opportunity. "Response" was the action taken by the same group of people to cope with the new situation. A challenge would arise as the result of many things – population growth, exhaustion of a vital resource, climate change. It was something that nobody had knowingly created. Response required vision, leadership, and action to overcome the threat and create a basis for survival and, hopefully, prosperity.

Because he analyzed large civilizations, Toynbee reserved the terms "challenge and response" for major threats and actions that impacted the well-being of the entire population. "Challenge" threatened the very survival of the existing system. "Response" would range from inaction to major change in the living conditions of individuals as well as the group. It could embody new technology, social organization, and economic activities, or a combination of various factors. "Response" was never predictable, and its outcome would only be known over time. This was the risk humans took – resulting in success or failure.

One of the examples used by Toynbee to describe the rise and fall of civilizations is the emergence of agriculture and cities in the ancient Near East. The challenge, in this case, was a regional shift in rainfall patterns.

North Africa, Egypt and Mesopotamia were no longer tracked by Atlantic storms which, for unknown reasons, moved further north. With less rain the traditional lifestyle of hunters and gatherers in this region could no longer be supported. Several response strategies emerged. Some people did nothing. They held on to their old ways, and eventually perished. Others migrated to find more amicable climatic conditions, and remained hunters and gatherers. But a few people survived, and even prospered, in the new environment by "inventing" the domestication of plants and animals, irrigated agriculture, and cities – the civilizations of Egypt and Sumer were born.

Critics of Toynbee have pointed out that the story is based on now dated archaeological evidence. Therefore, the chain of events leading to agriculture and urbanization may well have been different. Yet, Toynbee's categories of "Challenge and Response" are useful for understanding social change. They allow us to focus on important dimensions of change that will help us understand today's challenges.

First, major challenges arise infrequently. They are driven by transformation in environment, technology, economy, and society. The industrial revolution was one such challenge. Karl Polanyi spoke of "The Great Transformation" from agrarian to industrial society. During the industrial age three large transformations have taken place. None was as revolutionary as the end of the agrarian age, but each was powerful enough to spawn wars, revolutions, and massive social dislocations. And each eventually succeeded in creating jobs and livelihoods for larger numbers of people. The first hundred years of the transition was focused on access to raw materials. The next stage was dominated by the manufacture of finished goods. And now the world is undergoing another round of massive change through the global powers of knowledge and information.

Transformations of this magnitude break with the past in ways that go far beyond the normal process of change that occurs from generation to generation. It is understandable, therefore, that people feel threatened by change and want to resist it. Only a few have the vision to see necessity as well as opportunity in the midst of suffering and destruction.

Second, multiple responses are possible. All of them, including the option of inaction, carry risks and unknown outcomes. Not to act, at least for a while, actually seems more attractive to most, but eventually may claim the highest price.

Third, a successful response must be bold enough to overcome the threat and show a path to the new land. This requires strong leadership. Yet the leaders must also make the new vision acceptable to the majority of people. This condition is paramount in a democratic society. Unless a majority is found to support the new way, the solution will be unacceptable. Thus, the social process of transitioning from old to new conditions is critically important.

Finally, responses have a better chance of success if they allow for mid-course corrections. Large blueprints are inflexible and lead to social confrontation instead of bringing people together in pursuit of common goals.

Change in our time is closely associated with the emergence of a global economy dependent on rapid flows of information, technology, and capital. Benefits are large. But so are unintended consequences – economic, social and environmental. Economic globalization entails regional unemployment and new forms of endemic poverty.

Environmental globalization – the starting point for the debate in this volume – causes the most serious threats. Some are caused by the "Tragedy of the Commons". Garrett Hardin, in a famous essay written at the beginning of the environmental debate, used this phrase to warn against exploiting common property – grazing lands in the West, fish in the oceans, water in the river – without a keen eye on its sustainable yield and carrying capacity. Donella and Dennis Meadows, in *The Limits to Growth*, argued that population growth would outstrip the stocks of non-renewable resources, in particular fossil fuels. This particular prediction, following in the footsteps of Malthus and his concern about insufficient food for growing populations, was premature. New technologies, new discoveries and improved efficiency in using resources helped alleviate resource constraints. But the book was a milestone for focusing vigorous debate on linkages between natural, economic, and environmental factors.

In recent years, the very success of human industry has emerged as a direct threat to the functioning of natural systems – water, air, and land – that make possible human life on this planet. These new challenges fill the headlines of the papers and fuel heated debates: global warming, ozone depletion, loss of biodiversity, acid deposition, desertification, overpopulation, and resource intensive consumption. Many of these new issues are truly global in their reach, even though impacts may differ from place to place. They also require scientific research to identify and measure what is going on, making it difficult for many people to grasp their importance and urgency. In each case, human activities, helped by ever more powerful technology and steadily increasing numbers of people, create conditions that may cause serious, perhaps irreversible harm to natural forces on which human life depends. The Global Change Research Act of 1990 offers this definition of the new challenge: "Changes in the global environment (including alterations in climate, land productivity, oceans or other water resources, atmospheric chemistry, and ecological systems) that may alter the capacity of the Earth to sustain life."

The search for novel response strategies to address these global challenges is underway on a world-wide scale. It is still in its infancy. Most initiatives use *sustainable development* as their goal. The term means different things to different people, but a common foundation was laid by the Brundtland report (United Nations World Commission on Environment and Development, 1987). This report, after reviewing the new global challenges, suggested that they would dominate the international agenda of the twenty-first century. The report then proposed sustainable development as the appropriate response. Sustainable development, in this view, must seek three goals: economic and environmental priorities must be balanced, short-term and longer-term costs and benefits must be considered, and the stark differences in income and access to resources between rich and poor countries must be diminished.

This volume is published more than a decade into the debate on sustainable development. It seeks to highlight the many dimensions of sustainable development – not just economic and environmental, but also spiritual and religious, corporate and social, scientific and political. This introductory chapter establishes the link between global change and sustainable development. Sustainable development is a broad enough concept to respond to economic, social, as well as environmental challenges. It offers a new way to think about development in a world that is becoming more and more interdependent. Global change, in its many manifestations, is often referred to in the essays that follow, but this is not a book on climate change, population growth, poverty or any of the other specific driving forces of global change. Sustainable development acknowledges the seriousness of these problems and seeks solutions to these economic, environmental, and social issues. Global change and sustainable development are this generation's challenge and response.

Two specific events stand behind this book. Four years ago, the National Academy of Sciences initiated the *Global Commons Project* – an analysis of development issues during the next 50 years and the role science can play in identifying, monitoring and solving these challenges. As part of this effort, the Houston Advanced Research Center (HARC) examines the role of the private sector in sustainable development – one of the central themes in this book.

In 1997, the editors organized the DeLange/Woodlands Conference on Sustainable Development at the Rice University campus in Houston at which first results of the *Global Commons Project*, along with many other papers on sustainable development, were presented. The Academy released a major report of its findings in 1998 and convene a meeting of the world's academies of science in 2000. The Academy's approach in these activities is unique in that it focuses on the next 50 years. During these years a massive challenge needs to be met – to feed, house, and educate a world population with twice as many people as today. This stark reality defines what must be done in the lifetime of our children. The increase in population, with attendant increases in production and consumption, are already programmed. They cannot be undone. How compatible will these critical decades be with the principles of sustainable development? Will it be possible to defer to the end of this transition period the construction of human societies no longer dependent on constant economic growth? This transition to sustainability, and the agenda for action during this period, were at the heart of the DeLange ♦ Woodlands Conference.

This volume is made up of nine chapters that were commissioned following the Houston conference. Each author was asked to highlight and develop main conference themes. As a group the essays show that sustainable development, a decade into the debate, has become a powerful concept for thinking about the future – by no means a clear blueprint for action but a meaningful guidepost for leaders from business, society and policy.

To illustrate this point and summarize the main themes discussed in the book, we conclude with excerpts from the 1997 conference statement (HARC and Rice University, 1997). The report used the image of the road map to describe what lies ahead.

The destination

Most demographers project a rise in world population from the current 5.7 billion to between 8 and 12 billion by the middle of the next century. World population will then stabilize, and eventually decline. To manage the transition to sustainable development, we must find urgent answers to this question:

How can the economic and social systems of the world provide food, energy, jobs, education, and other amenities for this much larger population in ways that successfully balance economic and ecological needs?

The single most demanding challenge of sustainable development during this period will be to provide for the basic needs of this doubled world population. This will require continued rapid economic growth. By some estimates, the world economy will have to grow by a factor of eight. This growth estimate is based on a model that provides for meeting minimal needs for all people. We use this model because we agree that poverty and environmental degradation are two sides of the same coin.

We envision a sustainable world that is characterized by a balance between this level of economic growth and by ecosystem viability. Such a balance will enable us to mitigate the threat of global warming and other environmental problems. Achieving this balance depends on our acceptance of the limits to the absorptive capacity of the biosphere. We do not always know the boundaries of these limits, but we do agree that they exist.

The direction

Ethics and leadership

We agree that the examination of the role of ethics and leadership in shaping a sustainable society has been neglected. Yet, a successful transition will be realized only by a society that maintains respect for the natural environment. This value system represents the supporting pillars and principles that guide our actions through the transition. It dictates the manner in which we will respond to the challenge. It helps make possible the many detailed changes that are needed in technology and policy. Conversely, without ethical direction, managing the transition may well fail.

Sustainable development must not be allowed to become the domain of one ideology or political philosophy. The leaders that emerge through the transition are those that show the concrete appeal of sustainable development to widely diverse constituencies – developing countries as well as industrialized ones, industry as well as government, and loggers and fishermen as well as environmentalists. The concepts of ethics and leadership are difficult to define and analyze. It may seem implausible to consider changing an entire society's notions of these concepts and how they impact the transition. But these difficulties do not justify continued avoidance of discussing how ethics and leadership must play a role in the transition to sustainable development.

The scientific community and decision-making under uncertainty
Human actions are, among other factors, conditioned by our understanding of how the natural world works. Thus, science and technology play critical roles in informing decision-makers and citizens of the impacts of human activities on the behavior of the earth's ecosystems.

Public understanding of new discoveries by the scientific community is critical. Scientists must effectively communicate not just knowledge but also an understanding of and confidence in how that knowledge was acquired. Citizens for their part should not expect definitive answers in all cases, but they must be assured that scientific information is presented in good faith. Science can "tell us what will happen," "tell us the probability of its happening," and "tell us how we know." Citizens will choose how this information is used in the context of social, cultural, economic, and political knowledge or values.

At times, society does not know how to utilize scientific information in decision-making processes as it is often deemed incomplete and inconclusive. We sometimes have difficulty deciphering the signals we receive and hesitate to act decisively and effectively on this information. Accepting that our knowledge is limited and provisional, and that there may be limits to what we can know and predict, we must also accept the need to act on the current state of knowledge in our possession. We agree that we cannot always wait for science to reach its final understanding of complex issues. Rather we must utilize our social learning to move rapidly through the transition to sustainable development.

Market tools for managing the transition
We recognize the numerous obstacles, in the form of market failures and policy failures, that stand in our path to sustainability. Economists have long shown that the market system does not account for certain economic problems, such as externalities, free riders, and natural resource valuation. One area that deserves more attention is policy failures.

Policymakers, in their pursuit to create a particular industry or attempt to alleviate poverty, often develop policies that may not accomplish either objective, while resulting in significant environmental degradation. The evaluation of these "economic development" policies should be more rigorous, with special attention given to their environmental impacts.

We agree that the use of market tools should be the emphasis for an efficient, successful transition to sustainability. The command-and-control policy paradigm that characterized the 1970s and 1980s in the developed countries should be revisited. When governments mandate processes and procedures the focus remains on compliance rather than innovation. Regulation should be instituted by goals instead of by compliance methods. This flexibility would allow corporations to determine the best technical solutions to mitigate their own environmental impacts, thus minimizing their cost of achieving sustainability.

Well-designed market tools for sustainability contain within them a better and more efficient means of providing the information that producers, consumers, and decision makers require in order to make sound judgments that will ensure the sustainable future of our planet. Performance standards such as the ISO 14000 may help provide this type of information. Tools such as full cost accounting and life-cycle analysis can further promote sustainable practices by bringing to the forefront the full range of environmental costs associated with the production, distribution, and consumption of a product or service. Cost shifting tendencies should be reduced and prices should reflect marginal social costs, thereby providing appropriate and correct signals to decision makers in the economic and political process.

Stakeholders and dispute resolution

Development over the last few decades has led to dramatic improvements in living conditions in many countries, while considerable social and economic dislocation has occurred in other places. Historically, the poor have been denied a voice in decisions that directly and indirectly affect their futures. Consequently, they may have also been denied an equitable share in the distribution of resources and gains from economic activity.

The social dimension of sustainable development requires that the most directly affected people take a leading position in development initiatives. Stakeholder participation in decision making establishes a sense of ownership of all interested parties in regard to a specific action, be it a program, a project, or a legal act. It provides for consensus building, and hence, political sustainability of decisions that affect the lives and interests of different people and entities.

We agree that dispute resolution among competing stakeholders, business, and government is critical to sustainability. It is in the best interest of the project developers to allow for stakeholder participation in the case of local development projects. Well-planned stakeholder involvement will only enhance the possibilities for the successful completion of a project or endeavor. Conflict is costly and inefficient.

Effective dispute resolution, however, requires that the playing field first be leveled through the empowerment of local populations. Empowerment is obtained through education on legal and economic rights and the management of stakeholder participation.

Bringing it together

As we journey on the road to sustainability, certain difficult technical questions must be resolved. For example, how much growth can we enjoy, and of what kind, before environmental harm becomes unacceptable? How can we reduce growth in population and consumption? How will the increased human presence affect other living creatures? How can we ensure well-being for a larger portion of the population under the new growth structures? To these concerns, we have, currently, no definitive answers, but we have shown our points of agreement. Given these limits, the critical challenge is: Can we move forward into action or will we be trapped in endless debate?

We do not advocate a full blown blueprint for change that will outline in detailed steps all that needs to be done to ensure a sustainable Earth. The process of discovering these individual steps cannot be coordinated into a well-packaged whole. It is inherently an incremental process and we learn step-by-step by doing.

We feel optimistic about the future. Human creativity thrives on challenge, and we are confident that solutions will be found. But it will require – beyond ethical changes, beyond voluntary measures, and beyond technical fixes – the creation of a shared vision of a sustainable and desirable world. With a common language on sustainability we can develop a broad understanding so that individual and interdisciplinary action is part of a synergistic and comprehensive approach to sustainability that leverages all sectors of society. We stress again the urgency in getting started on the road to sustainable development.

REFERENCES

Hardin, G. 1968. "The tragedy of the commons." *Science* 162: 1243–48.
Houston Advanced Research Center and Rice University 1997. *Report of the 1997 De Lange/Woodlands Conference on Sustainable Development: Managing the Transition*. HARC, The Woodlands, Texas.
Meadows, D. H. & Meadows, D. L. 1972. *The Limits to Growth. A Report for the Club of Rome's Project on the Predicament of Mankind*. Universe Books, New York.
Polanyi, K. 1944. *The Great Transformation*. Farrar & Rinehart, New York.
Toynbee, A. J. 1946. *A Study of History Abridgement*. Oxford University Press, London.
United Nations World Commission on Environment and Development. 1987. *Our Common Future*. Oxford University Press, Oxford and New York.

Malcolm Gillis *and*
Jeffrey R. Vincent

2

National self-interest in the pursuit of
sustainable development

There is no universally accepted definition of sustainable development,[1] nor do all definitions of sustainable development yield practical guidelines for policymakers. The concept is perhaps best defined as development that maximizes the long-term *net* benefits to humankind, taking into account the costs of environmental degradation.[2] Net benefits include not merely income gains and reduced unemployment and poverty, but also healthier living conditions and other benefits associated with improved environmental quality. Interpreted this way, sustainable development stresses not the need to limit economic growth, as some have argued (e.g., Daly 1991), but rather the need to grow and develop sensibly, to ensure that the benefits of development are long-lasting: that in the most general sense, people become better off over time.

Sustainable development represents an attempt to make conservation the handmaiden of development, while protecting the interests of future generations. Pragmatic concepts of sustainable development value environmental protection not for its own sake, but for its contribution to the welfare of present and future generations. A sustainable development strategy thus permits the providential depletion of natural resources and the intelligent utilization of the environment's waste assimilation services. One key condition for achieving sustainability is that natural resources and environmental services not be undervalued or underpriced – a condition that is frequently violated in practice, as we shall see.

International meetings have tended to emphasize a global perspective on sustainable development. Most notable in this regard is the 1992 "Earth Summit" (officially, the U.N. Conference on Environment and Development) in Rio de Janeiro. The global perspective is summarized in the catch phrase, "Think globally, act locally." Interpreted literally, this slogan implies that humankind should first identify global environmental problems threatening the entire planet, and then individuals, households, communities, and, by extension, the nation-states that comprise the global village should do their bit to address those problems. At the Earth Summit, the global perspective caused two issues to dominate discussions and media coverage: first, international agreements aimed at three environmental problems – climate change, biodiversity loss, and unsustainable forest management (linked to biodiversity loss) – with global consequences, and second, financial transfers from the rich North to the poor South to implement *Agenda 21*, the United Nations' global blueprint for sustainable development.

While it is natural for diplomats at international conferences to focus on matters regarding international negotiations, the high-profile attention paid to the purely global dimension of sustainable development has distorted the essence of the concept. To be sure, there are some issues in sustainable development that require international agreements. Management of the world's oceans and transboundary waters, transboundary air and water pollution (e.g., acid rain), and climate change are prime examples. But, in this chapter, we argue that, to an extent rarely recognized in international conclaves on the subject, sustainable development is fundamentally a matter of national self-interest: following a development path that maximizes *national* net benefits as defined earlier (i.e., ones that are long-term and net of environmental degradation). We focus on developing countries, where the challenge of sustainable development is the steepest and the stakes the highest, as these countries are home to most of humanity. We argue first that poverty and many forms of environmental degradation are intertwined, and that economic growth is necessary to address both problems. We present evidence refuting the claim that scarcity of natural resources will limit economic growth at the global level. That is, we reject the view that economic growth can occur in developing countries only if economic activity is curtailed in developed countries. The scarcity issues that are pertinent for poverty alleviation and sustainable development occur primarily at the local and national levels, not the global level.

Next, we argue that economic growth, while important, is not sufficient for achieving sustainable development, due to the failure of market decisions to internalize the costs of environmental degradation. Governments around the world have often not only failed to enact policies to address market failures, but they have also pursued policies that have unnecessarily worsened environmental degradation. We argue that the most serious forms of environmental degradation in developing countries – and, for that matter, developed countries – result from these self-inflicted causes, not from global factors beyond their control. This implies that, in principle, policymakers in developing countries already have at their disposal the means to reduce environmental degradation. More significantly, the existence of these market and policy failures implies that reducing environmental degradation is in developing countries' self-interest. Sustainable development is not a "bad" that developing countries should pursue only if the global community compensates them through international financial transfers.

There is, however, a legitimate case for international compensation when environmental externalities cross national borders. Greenhouse gas emissions and biodiversity losses are good examples of such problems. In the final section of the paper we note that the selfish pursuit of sustainable development can make a substantial contribution toward addressing even truly global environmental problems such as these.

Environmental degradation, poverty, and economic growth

While sustainable development is an important concept for all societies, poor people in developing countries are far more dependent on their soils, forests, rivers, and fisheries than are citizens of rich countries. The World Bank (1997b) has estimated that natural capital – the capitalized value of rents[3] and services from the natural environment, including mineral deposits, land, forests, and fisheries – accounts for 10–20 percent of the total wealth of developing countries, versus only 2–5 percent in developed countries.[4] Environmental degradation thus looms as a much larger threat to life, health, and welfare in developing countries.

If economic growth inevitably worsens environmental degradation, then two fundamental aspects of sustainable development, poverty alleviation and environmental improvement, would seem to be at odds. This discouraging prospect follows from the fact that economic growth is the factor most robustly associated with poverty alleviation. A recent study by Gallup *et al.* (1997) lends strong support to the "rising tide" hypothesis. Using a unique new data set on income distribution over time in 64 countries, the authors found that the incomes of the poor have grown just as rapidly on average as aggregate income (per capita GDP). They found that other factors, such as health and education, play only a minor role in explaining higher incomes for the poor once economic growth is taken into account. This does not mean that health and education are unimportant; they are among the fundamental determinants of economic growth.

Fortunately for low-income nations, economic growth does not necessarily imply increased levels of environmental degradation. On the contrary, poverty itself is the prime adversary of improved management of many natural resources. In the absence of economic growth, the rural poor in developing countries are locked into livelihoods based on the exploitation of natural resources through farming, fishing, fuelwood collecting, and so on. Poverty induces rapid depletion of these resources and reduces investments in conservation, as it is associated with high discount rates.[5] By creating a broader range of employment opportunities for the rural poor,[6] economic growth thus takes pressure off the resource base. By raising incomes, it contributes to a decrease in discount rates, thus encouraging patterns of resource use based on a longer time horizon.

Poverty as a cause of deforestation

There can be little doubt that poverty by itself or in combination with other factors is the main cause of deforestation in most developing nations. Consider the case of Ghana, where per capita GDP declined at an average annual rate of 2.1 percent from 1965 to 1983. In 1900, one-third of Ghana's land area was covered by virgin tropical moist forest. When one of us first worked in Ghana in 1969–71, such forest still covered about 20 percent of the land (Gillis 1988b); there was still a lot of forest to study. No more. By 1990, virgin forest cover had shrunk to less than 7 percent of Ghana's land area (Gillis 1991b). As elsewhere in sub-Saharan Africa, and also in South and Southeast Asia, Latin America, and the Caribbean, poverty is destroying the forest. The landless poor in Ghana, Madagascar, India, the Philippines, Ecuador, and El Salvador practice destructive slash-and-burn agriculture not because they are ignorant or venal, but because decades of slow or no economic growth have left them with no better alternatives. Nearly 1.5 billion people in the world live in absolute poverty; at least a third are landless poor engaged in destructive forms of shifting cultivation (Gillis 1997). "Solutions" to tropical deforestation that do not take into account the needs of the poor by raising their incomes are no solutions at all.

Slash-and-burn agriculture is not the only manifestation of the effects of poverty on deforestation. In many poor nations, it has been accompanied by the ever more desperate search for fuelwood by people lacking the means to climb up the energy ladder. In Ghana in the mid-1980s, for every cubic meter of wood harvested for industrial uses (sawnwood, plywood, etc.), eight cubic meters were cut for fuelwood (Gillis 1988b). The connection between poverty and deforestation in Haiti is so well known as to require no further comment (Repetto 1988). For developing nations generally, FAO (1997) estimated that 80 percent of the roundwood harvested in 1994 was fuelwood for cooking and other household uses, not logs for commercial wood products like sawnwood and plywood.

According to Myers (1994), world-wide the tropical forest estate shrank by about 142 000 km² in 1989. Most of the decrease, about three-fifths, was attributable to the activities of poor people, as about 87 000 km² fell to slash-and-burn agriculture. The balance was accounted for by unduly destructive logging methods, primarily in Southeast Asia (30 000 km²); cattle ranching, mostly in Brazil and Central America (15 000 km²); and tea, rubber, and oil palm plantations and dam, mining, and road building projects (10 000 km²). These figures are roughly comparable to ones for 1980–90 from FAO's 1980 and 1990 Forest Resources Assessments (FAO 1997), which estimated that shifting cultivation accounted for 46 percent of the loss in closed forest cover and that permanent agriculture, cattle ranching, and reservoirs and other infrastructure projects accounted for another 40 percent.

Given the link to poverty, it is not surprising that cross-country data indicate an inverse relationship between deforestation rates and per capita income. Figure 2.1 shows this relationship, based on data from the 1980 and 1990 FAO Assessments (as reported electronically in the World Bank's *STARS* database; World Bank 1993). On average, the deforestation rate dropped by 0.06 percentage points for each US$1000 increase in per capita GDP (1987 price levels). In most developed countries, the deforestation rate was actually *negative*: forest area increased.

While some might argue that rich countries have "exported" deforestation by shifting toward tropical forests as a source of roundwood for the forest products they consume, there is little evidence to support this argument (Vincent 1992). First, as noted above, commercial logging directly accounts for only about a fifth of tropical deforestation. While logging also indirectly contributes by improving access for shifting cultivators, most deforestation from shifting cultivation occurs in countries that are relatively insignificant tropical timber exporters. Second, very little of the harvest of industrial roundwood in developing countries, only 17 percent in 1970 and 5 percent in 1994, is exported to developed countries as logs, sawnwood, or wood-based panels.[7] Most is consumed domestically. Finally, while the harvest of industrial roundwood in developing countries increased by 209 million cubic meters during

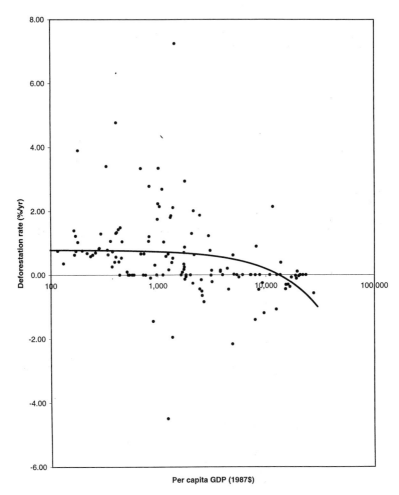

Figure 2.1 Deforestation rate (percent per year during 1980–90) vs. per capita GDP (average for 1980–89 in US$, 1987 price levels). Number of countries: 136. Regression line: deforestation rate = 0.786 – 0.0000596 per capita GDP. Number of observations = 136, R^2 = 0.081; t-statistic for coefficient on per capita GDP = 3.43.

1970–94, consumption of sawnwood and wood-based panels in developed countries actually *decreased* by 51 million roundwood equivalents.[8]

Far from "exporting" deforestation, developed countries have instead stabilized their forest areas precisely because they have *developed*: their agricultural frontiers closed as a side effect of the growth of nonagricultural sectors. This phenomenon can already be observed in some of the most rapidly growing developing countries. Peninsular Malaysia had one of the most rapid deforestation rates in the developing world in the late 1970s, 1.2 percent per year (FAO/UNEP 1981). This compared to an average of 0.6 percent per year for tropical Asia as a whole. By the early 1990s, however, the deforestation rate in Peninsular Malaysia had fallen to negligible levels. Rapid industrialization, led by labor-intensive, export-oriented manufacturing, had induced rural-to-urban migration, raised the opportunity cost of rural labor, and reduced the returns to forest conversion (Vincent and Yusuf 1993; Vincent *et al.* 1997b Chapters 4 and 5). For such reasons, economic growth plays a role in reducing deforestation stemming from not only subsistence activities such as shifting cultivation and fuelwood collection but also commercial agricultural activities such as tree-crop plantations and ranches.

The negative correlation between pollution and income

Deforestation is not the only form of environmental degradation that is negatively correlated with income. In the 1970s, in the wake of the Stockholm Conference on the Environment (the original "earth summit"), the World Health Organization and the United Nations Environment Program established the Global Environmental Monitoring System (GEMS). This system includes air and water quality monitoring stations located in dozens of developing and developed countries around the world. Figures 2.2–2.7 show the simple association between per capita GDP (1987 price levels) and ambient concentrations of two air quality parameters – total suspended particulates (TSP) and sulfur dioxide (SO_2) – and four water quality parameters – dissolved oxygen, nitrogen (nitrate + nitrite), total suspended solids, and fecal coliform.[9] These are key indicators of pollution, and they are the ones with the most abundant data in the GEMS database. With the exception of dissolved oxygen, higher levels of all six parameters indicate worse pollution.

Although there is much scatter in the figures, in each case higher income is, on average, associated with improved environmental quality. The figures include the regression line for this average relationship, determined by regressing the natural logarithm of the air or water parameter in question on a constant and the natural logarithm of per capita GDP. For all parameters except nitrogen, the coefficient estimates for per capita GDP were statistically significant at the 1 percent level.[10] The regression coefficients imply that a 10 percent increase in per capita

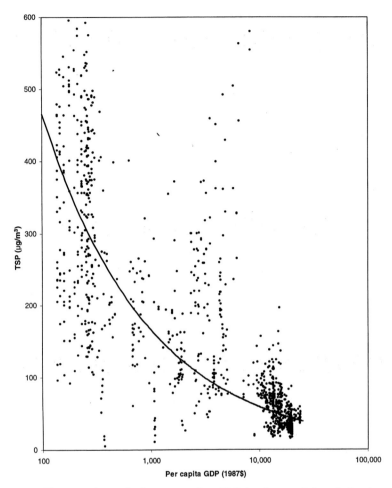

Figure 2.2 Atmospheric concentration of TSP (total suspended particulates) vs. per capita GDP (US$, 1987 price levels). Number of countries: 28 (multiple monitoring stations in some countries), time period was 1972–93 (incomplete time series for some stations). Regression line: ln(TSP concentration)=8.20 − 0.447 ln(per capita GDP). Number of observations=1307, R^2=0.698; t-statistic for coefficient on ln(per capita GDP)=54.9.

GDP is associated with a 4.5 percent decrease in TSP and a 0.6 percent decrease in SO_2 in the air, and, in surface waters, a 1.3 percent increase in dissolved oxygen (which, recall, signals environmental improvement), a 0.3 percent decrease in nitrogen, a 5.0 percent decrease in suspended solids, and a 2.8 percent decrease in fecal coliform.

These results demonstrate that tradeoffs between environmental

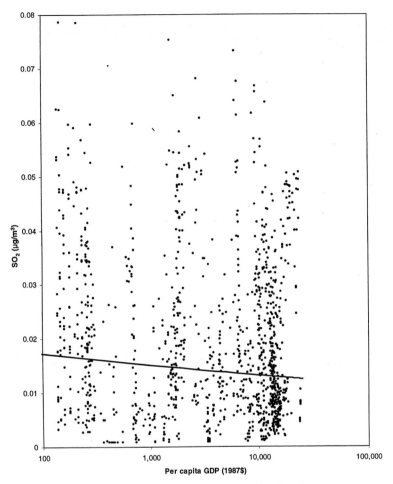

Figure 2.3 Atmospheric concentration of SO₂ (sulfur dioxide) vs. per capita GDP (US\$, 1987 price levels). Number of countries: 40 (multiple monitoring stations in some countries), time period was 1971–93 (incomplete time series for some stations). Regression line: ln(SO₂ concentration)=3.79−0.0585 ln(per capita GDP). Number of observations: 1252, R^2=0.010; t-statistic for coefficient on ln(per capita GDP)=3.50.

improvement and income growth are not inevitable. Environmental quality can improve even if economic activity rises. This is not to say, however, that economic growth will *automatically* lead to improved environmental quality (Arrow *et al.* 1995). That is, the negative correlations in Figures 2.2–2.7 between pollution and per capita GDP should not be interpreted as evidence of a causal mechanism. Although the general

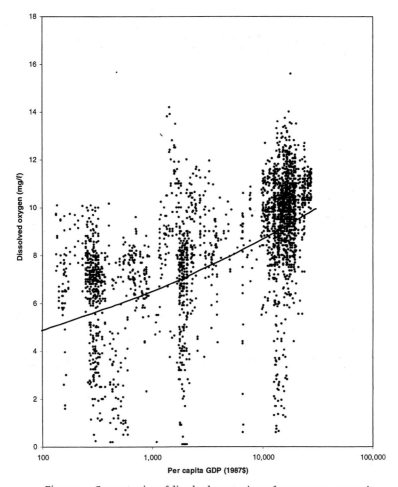

Figure 2.4 Concentration of dissolved oxygen in surface waters vs. per capita GDP (US$, 1987 price levels). Number of countries: 54 (multiple monitoring stations in some countries), time period was 1979–94 (incomplete time series for some stations). Regression line: ln(dissolved oxygen concentration)=1.01+ 0.125 ln(per capita GDP). Number of observations=2682, R^2=0.172; t-statistic for coefficient on ln(per capita GDP)=23.6.

tendency is clear and statistically significant, the scatter in the figures is such that environmental quality is higher in some developing countries than in some developed countries.[11]

There are, however, reasons to expect that economic growth facilitates improvements in air and water quality: that it is a necessary, if not a sufficient, condition. One is structural change. At higher incomes, the struc-

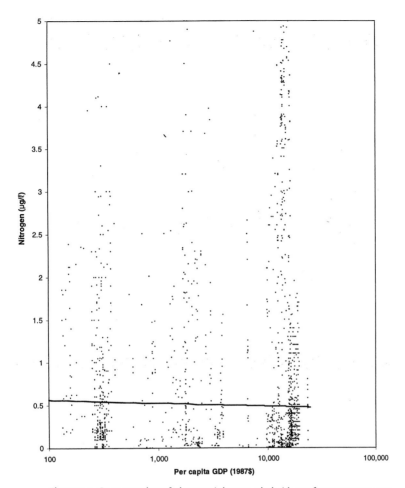

Figure 2.5 Concentration of nitrogen (nitrate + nitrite) in surface waters vs. per capita GDP (US$, 1987 price levels). Number of countries: 45 (multiple monitoring stations in some countries), time period was 1979–94 (incomplete time series for some stations). Regression line: ln(nitrogen concentration) = 0.436 – 0.0300 ln(per capita GDP). Number of observations: 1671, $R^2 = 0.001$; t-statistic for coefficient on ln(per capita GDP) = 1.17.

ture of the economy tends to shift away from heavy industry toward less polluting light manufacturing and service sectors. On the other hand, at somewhat lower incomes the shift from agriculture to manufacturing that reduces deforestation also increases the generation of industrial pollutants, so the impact of structural change is not necessarily salutary at all income levels.

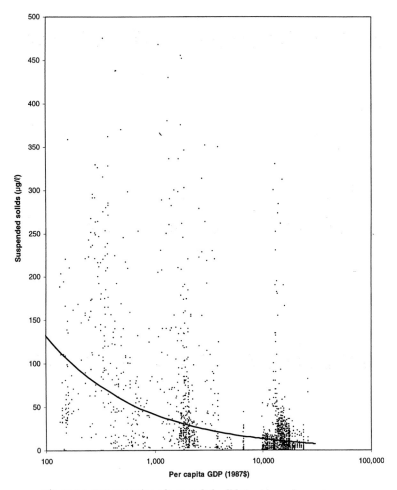

Figure 2.6   Concentration of suspended solids (105) in surface waters vs. per capita GDP (US$, 1987 price levels). Number of countries: 42 (multiple monitoring stations in some countries), time period was 1979–94 (incomplete time series for some stations). Regression line: ln(nitrogen concentration) = 7.20 – 0.503 ln(per capita GDP). Number of observations: 1700, R^2 = 0.223; t-statistic for coefficient on ln(per capita GDP) = 22.1.

Another, less ambiguous reason is rising demand for environmental quality: As incomes rise, people become more willing to pay for environmental improvement.[12] Economic growth also makes people, and countries, better *able* to pay for environmental protection. Departments of environment are routinely the most underfunded and understaffed government agencies in developing countries. While environmentalists

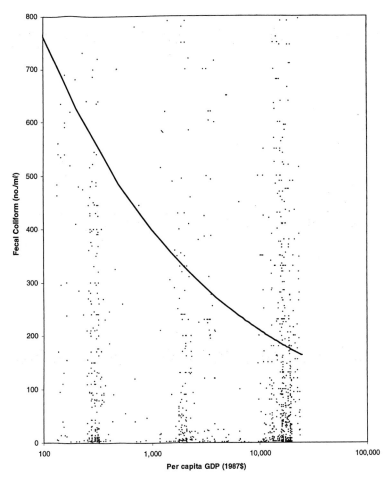

Figure 2.7 Concentration of fecal coliform bacteria in surface waters vs. per capita GDP (US$, 1987 price levels). Number of countries: 43 (multiple monitoring stations in some countries), time period was 1979–94 (incomplete time series for some stations). Regression line: ln(fecal coliform concentration) =7.92−0.279 ln(per capita GDP). Number of observations: 1736, R^2=0.022; t-statistic for coefficient on ln(per capita GDP)=6.22.

often criticize economic reform programs for slashing already meager environmental budgets (Reed 1992), it is hard to imagine how higher environmental budgets can be achieved and sustained in the absence of economic growth stimulated by reform programs.

We emphasize again that we are not arguing that economic growth automatically leads to environmental improvement. Economic growth is an essential, but not the only, ingredient. By raising the demand for environmental quality and the supply of financial resources to meet that demand, it sets the stage for more deliberate policy responses to address environmental problems. We discuss those responses later in this chapter.

Global abundance of nonrenewable resources

Can developing countries hope to achieve the combination of high incomes, stable or expanding forest areas, and generally low pollution levels found in the developed world? The nay-sayers claim that the earth's limited supply of natural resources makes economic growth a zero-sum game. This is an old argument, going back at least to Malthus in the late eighteenth century. The Club of Rome study, *The Limits to Growth* (Meadows *et al.* 1972), revived it in the early 1970s. The sequel to that study, *Beyond the Limits* (Meadows *et al.* 1992), kept it in the news (and attracted a strong rebuttal; see Nordhaus 1992). Most recently, this strongly pessimistic contention has been cast in terms of differences in material consumption levels between rich and poor countries: that the planet does not have sufficient resources, in particular nonrenewable ones, to enable developing countries to achieve the standard of living of developed countries unless the latter reduce their consumption levels.[13]

The available evidence does not support this contention. Perhaps the most dramatic projections in *The Limits to Growth* were the estimates of the number of years until reserves of nonrenewable resources would be exhausted. That study made its projections by dividing estimates of global reserves (identified resources that can be profitably exploited at current prices and technology) by annual global mine output. It adjusted the latter for projected future output growth, set equal to post-war trends. According to such projections, gold should have been exhausted by 1979, silver and mercury by 1983, tin by 1985, zinc by 1988, petroleum by 1990, copper and lead by 1991, and natural gas by 1992.

These projections were obviously wrong. They were so far off the mark because they ignored several factors that neo-Malthusians unfortunately continue to downplay. One is the ongoing growth of reserves due to discoveries and improved technology. The second column in Table 2.1 shows that reserves of most major minerals either increased during 1975–95 (coal, petroleum, bauxite, iron ore) or fell much less rapidly than one would have expected given the amounts produced during the period (copper, nickel, phosphate rock, potash). Reserves increased particularly rapidly for fossil fuels, which neo-Malthusians pay particular attention to

Table 2.1. *Global reserves and production of nonrenewable resources*

	Trend in reserves[a] 1975–95[b]	Ratio: Identified resources/Reserves 1995[b]	Trend in production[a]	
			1960–74	1974–95[b]
	%/year		%/year	%/year
Fossil fuels				
Coal	+3.0	7.9	+1.2	+1.8
Petroleum	+2.0	n.a.	+7.3	+0.8
Metals				
Bauxite	+1.4	2.8	+7.5	+1.6
Copper	−1.4	7.4	+4.0	+1.4
Iron ore	+2.6	5.3	+3.9	+0.5
Nickel	−0.7	2.8	+6.4	+0.7
Fertilizer components				
Phosphate rock	−1.9	4.7	+6.5	+1.5
Potash	−0.9	29.8	+6.9	+0.5

Notes:

[a] Trends are based on point-to-point estimates, not logarithmic regressions.

[b] For fossil fuels, data are for 1993, not 1995.

due to the fundamental importance of energy to economic activity. And the table does not even include natural gas, which the Energy Information Administration (1997) projects will surpass coal as a primary fuel by 2015, thanks to huge discoveries during the past 25 years. World natural gas reserves rose from 2500 trillion cubic feet in 1975 to 4945 trillion cubic feet in 1997.

While future increases in reserves are not guaranteed, the pace of discovery has accelerated for key minerals such as petroleum (recent discoveries in the Caspian basin rivaling reserves in the Persian Gulf) and coal (substantial discoveries in Indonesia, Colombia, and Venezuela since the 1980s). Proven oil reserves are 30 percent higher today than in 1970 (Energy Information Administration 1997). Estimates of undiscovered and ultimately recoverable conventional oil resources range up to 1 trillion barrels, which is nearly 50 percent larger than the total cumulative amount of oil produced up to the present day (Energy Information Administration 1997).

Even without additional discoveries, substantial physical deposits have been identified that can move into the reserve category as technology improves or, if technology does not improve, prices rise in response to scarcity. The third column of Table 2.1 shows that the resource/reserve ratio is especially high for most of the minerals whose reserves declined during 1975–95, especially copper and potash. During 1985–94, technological advances caused productivity in coal mining to rise by 5–9.6 percent per year in Australia, Canada, South Africa, and the U.S.A. (Energy Information Administration 1997). Even with today's technology, a trillion barrels of oil can be recovered from nonconventional resources like tar sands and oil shale if the price of oil rises to US$40 per barrel (Energy Information Administration 1997). In 25 years, improved technology is projected to raise the recoverable amount at that price to 5 trillion barrels. Technology is also markedly improving oil recovery in offshore locations.

The Limits to Growth also ignored several forces that have dampened the rate of growth in global consumption of nonrenewable resources. The fourth and fifth columns of Table 2.1 compare average annual growth rates in global mine output during 1960–74 and 1974–95. With the sole exception of coal, which was the physically most abundant resource ca. 1970 according to The Limits to Growth (111 years until exhaustion; much longer now, thanks to higher reserves), mine output grew much less rapidly during the second period. What explains this sharp change?

To begin, world population has not grown nearly as rapidly as demographers forecast in the 1960s and 1970s (Cohen 1995). The population growth rate continues to decline more rapidly than projected, as the "demographic transition" (the fall in fertility rates associated with economic development) has been much stronger than expected. Indeed, some observers have begun warning of problems that could arise from too *slow* population growth, such as an inadequate work force to support those who have retired.[14]

But demography cannot be the entire explanation, as all the growth rates for global mine output including coal during 1974–95 were at or below the global population growth rate. Moreover, global GDP doubled during the same period. The rest of the explanation is that the consumption of goods and services made from nonrenewable resources is not strictly proportional to either population or income. Energy consumption illustrates this point particularly clearly. Figure 2.8 shows the relationship between per capita total energy consumption (kilograms of oil

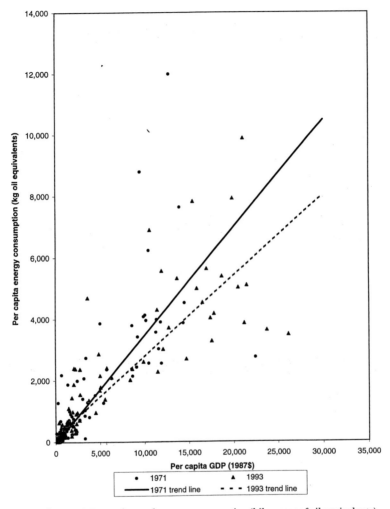

Figure 2.8 Per capita total energy consumption (kilograms of oil equivalents) vs. per capita GDP (US$, 1987 price levels) across developing and developed countries in 1971 and 1993. Number of countries: 110. Equations of regression lines: for 1971, per capita energy consumption = 126 + 0.351 per capita GDP (R^2 = 0.763; t-statistic for coefficient on per capita GDP = 13.5); for 1993, per capita energy consumption = 283 + 0.262 per capita GDP (R^2 = 0.867; t-statistic for coefficient on per capita GDP = 17.4).

equivalents) and per capita GDP (1987 U.S. dollars) in the 110 countries with data in both 1971 and 1993 (the earliest and most recent years with good representation for both developing and developed countries). Within each year, the relationship was approximately proportional: Countries with twice the per capita GDP had twice the per capita energy consumption. But during 1971–93, the relationship shifted downward strongly and became less steep, with per capita energy consumption at any given income level falling by a quarter. A US$1000 increase in per capita GDP was associated with increased energy consumption on the order of 351 kilograms of oil equivalents in 1971 but only 262 kilograms in 1993.

These changes, and the decrease in growth rates between the fourth and fifth columns of Table 2.1, reflect behavioral responses to the "oil shocks" of the 1970s, which contributed to price increases for a broad range of minerals, not just oil. The price increases prompted a tremendous conservation response, as industries and households began using raw materials and energy more efficiently, recycling more, and substituting alternative materials (e.g., fiber optics for copper wire). Figure 2.9 shows the dramatic de-linking of energy consumption and income that has occurred in the U.S.A. since the early 1970s. The relationship between per capita energy consumption and per capita GDP was proportional until 1973, the year of the first "oil shock." During the subsequent 20 years, per capita GDP increased by more than 40 percent, while per capita energy consumption showed no overall trend, fluctuating within 5–10 percent of its 1973 value.

Higher prices are not good news for resource-importing developing countries. Although Table 2.1 indicates that the huge amounts of resources consumed by developed countries since the onset of the Industrial Revolution have not left the table bare, resource-importing developing countries will be worse off if prices increase just as these countries are reaching for their share. But this has not happened, and it is unlikely to happen anytime soon. The price increases of the 1970s were transient blips on a relentlessly downward trajectory of resource prices. Nordhaus (1992) presented long-term data (1870–1989) on the real prices of ten minerals: petroleum, coal, zinc, copper, lead, iron ore, aluminum, phosphorus, molybdenum, and sulfur. He calculated real prices by dividing the world market price for a unit of an extracted mineral by hourly manufacturing earnings in the U.S.A. He found that real prices have fallen dramatically for all ten minerals since the late 1800s. For example, the real price of a barrel of oil in 1989 was only about a fifth of the real price in

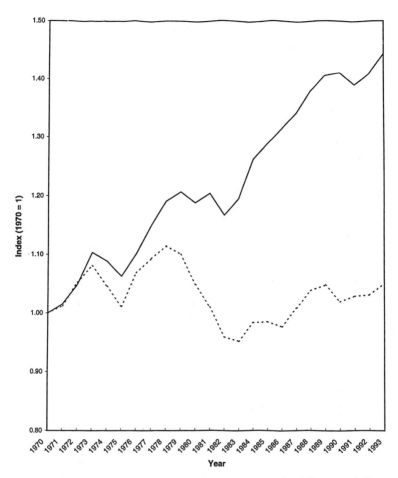

Figure 2.9 Trends in per capita total energy consumption (kilograms of oil equivalents) and per capita GDP (US$, 1987 price levels) in the United States during 1970–93.

1870. This means that the average worker in the U.S.A. only had to work one-fifth as long in 1989 to obtain the money needed to purchase a barrel of oil or the oil content of goods made from it.

Have resources become similarly cheaper to developing countries? Data on earnings are readily available for developing countries only since the early 1960s. Table 2.2 shows trends in real resource prices from then until the early 1990s in the largest countries with the most complete data in the three principal developing regions (Africa, Asia, Latin America). It also shows real price trends in South Korea, one of the fast-growing (until

Table 2.2. *Trends in real natural resource prices (deflated by annual manufacturing earnings per employee), early 1960s to early 1990s*

Calculated by regressing natural logarithms of annual values on a time trend and a constant. Numerical values are shown only for values significantly different from zero at the 5 percent level; n.s., not significantly different from zero. Sample period: Brazil, 1963–91; India and Kenya, 1963–92; South Korea, 1963–93.

Resource	Brazil	India	Kenya	South Korea
	%/year	%/year	%/year	%/year
Fossil fuels				
Coal	n.s.	n.s.	+3.5	−8.2
Petroleum	n.s.	+4.8	+5.8	−5.6
Metals				
Aluminum	−2.9	n.s.	n.s.	−10.1
Copper	−4.0	n.s.	n.s.	−12.2
Iron ore	−4.3	−1.5	n.s.	−11.0
Nickel	n.s.	n.s.	+2.2	−9.1
Fertilizer components				
Phosphate	−2.4	n.s.	n.s.	−9.6
Potash	−2.2	n.s.	+2.0	−8.6

recently) East Asian "tigers." In Brazil and India, real prices either declined or remained constant in all cases but one (petroleum in India). In South Korea, real prices declined at astonishingly rapid rates. Only in Kenya were several resources, four out of eight, more costly in the early 1990s than in the early 1960s. This is a direct consequence of Kenya's minimal economic growth since independence.

The important story concerning economic growth and nonrenewable resources is therefore not that physically limited resources will constrain growth at the global level, but rather that resources will become economically more scarce in countries that grow slowly. The solution to resource scarcity is more growth, not less. Thanks to discoveries and improved technologies, nonrenewable resources are more available to developing countries today in both physical and, more importantly, economic terms than they were to developed countries when the latter began industrializing more than a century ago.

The consequences of resource depletion are also less threatening today than they have ever been, as the costs of backstop technologies have continued to decline. Consider the case of renewable energy sources that can substitute for fossil fuels. The cost of electricity generated by wind-power fell from about US$1 per kilowatt-hour in the 1970s to 5 cents in 1996 (Energy Information Administration 1997). The cost of photovoltaic cells fell from 89 cents per kilowatt-hour in 1980 to 30 cents in 1990, and it is projected to decline to only 15 cents by 2000 (Asian Development Bank 1997, Table 4.7). Until recently, powering a small automobile with proton-exchange-membrane fuel cells cost some US$30 000 for just the platinum in the cells (Anonymous 1997). Thanks to technological improvements, only $140 worth of platinum is now needed. The cost of power generators based on phosphoric-acid fuel cells is now only double the cost of conventional generators – it has fallen by half in just the past two years – and it is likely to fall further with mass production (Anonymous 1997).

Depletion and growth: a national, not global, issue

Those who persist in seeing signs of gloom and doom might argue that, since many developing countries are primary commodity exporters, declining resource prices are hardly a reason to rejoice. A "resource curse" from declining terms-of-trade (damned if prices fall) and other factors (e.g., "Dutch disease" if commodity markets boom – damned if prices rise) is a familiar story to development economists, and there does seem to be something to it. Sachs and Warner (1995) found that per capita GDP grew less rapidly during 1971–89 in resource-rich countries than in resource-poor countries after controlling for other factors affecting their growth rates.

But not all resource-rich countries have suffered the curse. In Southeast Asia, resource-rich countries such as Indonesia and Malaysia were among the fastest growing economies in the world from the 1960s through 1996 despite facing the same declining resource prices as other developing countries, despite depleting their own resources very rapidly, and despite serious, long-standing problems in the banking sector that finally came to a head in the fall of 1997 (Gillis 1998). What factors distinguished Indonesia and Malaysia from most other resource-rich developing countries after 1960? Generally sound macroeconomic policies – in particular, open economies – are the most important ones (Asian Development Bank 1997). These policies created a favorable environment for public and private investment in human capital (education, skills, technology) and physical capital (equipment, structures, infrastructure). This is a critical point, as the most fundamental condition for economic sustainability is that a country must maintain the value of its total capital stock (total wealth): the sum of its human, physical, and natural capital stocks (Solow 1986). Raising the standard of living requires increasing total wealth, which means that net investment, the sum of additions to all capital stocks minus depreciation of those stocks, must be positive.

High rates of investment in human and physical capital enabled Indonesia and Malaysia to offset the depletion of their natural resources and to increase their total wealth. Repetto *et al.* (1989) determined that cumulative net investment in Indonesia was positive during 1971–84 after deducting the depreciation of natural resources (petroleum, timber, agricultural soils) from gross investment. Technical corrections to the depreciation allowance for petroleum suggest that net investment was even higher than Repetto *et al.* estimated (Vincent *et al.* 1997a). Similarly, Vincent (1997; see also Vincent *et al.* 1997b, Chapter 2) found that net investment was positive in Malaysia during 1970–90 after deducting the depreciation of physical capital and two important types of natural capital (minerals and timber).

The Malaysian case is instructive at the subnational as well as the national level, as it contains contemporaneous examples of both sustainable and unsustainable development. Peninsular Malaysia is separated from the states of Sabah and Sarawak on the island of Borneo by nearly 400 miles of the South China Sea. After an inauspicious, largely wasteful start earlier this century, Peninsular Malaysia enjoyed sustainable development for nearly three decades after 1965, successfully capitalizing (literally) upon its natural resource base. Net investment adjusted to reflect resource depletion was positive during every year from 1970 to 1990. Real economic growth (per capita GDP) was in excess of 3 percent during 1965–90 and nearly 6 percent since. By the mid-1990s, this rapid growth had virtually banished rural poverty as a cause of deforestation.

Sabah and Sarawak originally had, if anything, an even richer natural resource base than Peninsular Malaysia. But for Sarawak since the mid-1980s and Sabah since 1970, development has been unsustainable, with net investment rates falling below zero. Development in both states continues to be plagued by the scourge of rural poverty, and the economies of the two states have weakened since the early 1990s. While Peninsular Malaysia has made a highly successful economic transformation, building up reproducible capital stocks as it "cashed in" its natural capital, Sabah and more recently Sarawak have in effect been consuming wealth rather than accumulating it.

The message from Indonesia and Malaysia is that what determines long-run economic growth prospects in resource-rich developing countries is how a country manages its economy and uses the earnings from resource extraction – not international market conditions beyond its control. By and large, Indonesia and Malaysia managed their economies well during the three decades prior to the current financial crisis. The same factors that were responsible for their strong performance until mid-1997 make it highly likely that both nations will surmount their recent difficulties. Certainly both have directly or indirectly reinvested much of their resource earnings. Resource-rich developing countries that have failed to grow have typically not done this. The World Bank (1997a) estimated that net investment ("genuine savings") was negative during most of 1970–93 in sub-Saharan Africa, the economically most stagnant region of the world. Net investment fluctuated between positive and negative in Latin America, where some growth did occur, and it was strongly positive in fast-growing East Asia and the Pacific.

Degradation of agricultural soils and depletion of freshwater

Estimates of net investment provide accurate measures of changes in the total capital stock only if they include renewable as well as nonrenewable resources. The studies discussed above included renewable resources only to a limited degree. For example, the study of Malaysia by Vincent (1997) and the cross-country study by the World Bank (1997a) included the depreciation of timber stocks, and the study of Indonesia by Repetto *et al.* (1989) included the depreciation of agricultural soils as well. Another study by Repetto *et al.* (1991) of Costa Rica, added coastal fisheries to the list. But most net investment studies have not covered renewable resources very comprehensively.[15] We are not aware of *any* studies that have included plausible estimates of the depreciation of freshwater resources or air and water quality.

Developing countries' greater dependence on a more degraded resource base makes the scanty treatment of renewable resources more consequential for them than for developed countries. Furthermore, they, not the world as a whole, stand to lose the most from ignoring resource degradation. Two key renewable resources other than forests, agricultural soils, and freshwater, illustrate this point. Like deforestation, the degradation of agricultural soils and the depletion of freshwater resources tend to be developing country problems, not global problems.

World food experts agree that opportunities to increase food production by expanding cultivated area are very limited, especially in developing countries (Crosson 1997). This means that the increase in world food supplies needed to fight hunger and to feed the planet's growing population must come primarily from higher yields (higher crop output per hectare). Despite popular views to the contrary (e.g., Brown and Wolf 1984, Myers 1993), erosion, salinization, and other forms of soil degradation are not a critical world-wide constraint on yields. Crosson (1995, 1997) has noted that reliable data on soil degradation and its impacts on yields did not become available until the early 1990s. Combining estimates of the severity of soil degradation from Oldeman *et al.* (1990) with estimates of yield impacts from Dregne and Chou (1992), he calculated that the global impact of soil degradation during 1945–90 was equivalent to only 0.1–0.2 percent of the value of agricultural output. Similarly, Huang *et al.* (1997) projected that soil degradation will reduce grain production in China, the key country in the world food economy, by only 1 percent by 2020. In a pair of papers that were unique in analyzing soil data dating back to the 1930s, Lindert (1996a,b) argued that even less soil degradation has occurred in China than the estimates in Oldeman *et al.* (1990) indicate. He found that "total phosphorus and potassium have generally risen," the "topsoil layer has not gotten thinner," and declines in soil organic matter and nitrogen on cultivated lands have made "little difference," thanks to fertilizer applications. He reported similar findings for Indonesia.

Soil degradation is instead a local and regional problem, and it occurs primarily in developing countries. Crosson (1997) identified as key "hot spots" the Indus, Tigris, and Euphrates river basins, northeastern Thailand, steep areas in China and Southeast Asia, the Nile delta, portions of the Sahel and Nigeria, irrigated areas in northern Mexico, and steep areas in sub-humid and semi-arid Central America. Labeling soil degradation a "local and regional problem" is not intended to diminish its significance. Rather, it highlights the fact that developing countries, especially the poor and malnourished in developing countries, will bear the brunt of its consequences.

Inputs and technology, not soil degradation, are the critical factors affecting crop yields in developing countries. While there is considerable scope to raise yields by adjusting inputs toward more optimal levels (see Larson and Frisvold 1996 for a discussion of fertilizer use in Africa), in the long run the most important factor is improved technology. All experts emphasize the need to invest more in agricultural research to break through yield ceilings for key crops like rice and wheat (Crosson 1995, Rosegrant *et al.* 1995, Timmer 1997, Goldman, personal communication).[16]

The situation with regard to freshwater supplies is similar. The world's renewable freshwater supply, defined as the sum of runoff plus infiltration, was around 8000 m^3 per capita in 1990 (Gleick 1993a, Cohen 1995). Frequently cited estimates by hydrologist Malin Falkenmark give supplies of 1670 m^3 per capita and 1000 m^3 per capita as the thresholds for water stress and water scarcity, respectively (Cohen 1995, Serageldin 1995). Even a doubling of world population (thus, a halving of per capita supply) would leave per capita supply comfortably above these thresholds.

But this simple calculation ignores the tremendous variation in freshwater supplies that exists across regions. Water stressed and water scarce regions do of course exist, and they tend to be in developing countries with fast-growing populations. For example, although only 8 percent of the world's population lived in countries with less than 2000 m^3 of water per capita in 1990, 24 percent of the population in sub-Saharan Africa and 71 percent of the population in the Middle East and North Africa did (World Bank 1992). Income growth can be expected to increase the pressure on freshwater supplies in developing countries: per capita water withdrawals in the early 1990s were three times higher in high income countries than in low income countries (World Bank 1992).

Does this mean that a freshwater crisis is inevitable in developing countries? Fortunately, no. Several factors, analogous to those for deforestation and nonrenewable resources, can potentially mitigate the effects of water scarcity. The first is structural change: As income rises, water withdrawals shift away from irrigation (91 percent of withdrawals in low income countries) toward industrial and domestic (household) uses (61 percent in high income countries). Because irrigation's share is so large in developing countries, relatively modest allocations away from it can meet much of the increased demand by industries and households.

Second, there is great scope for more efficient use of water in both irrigation and other uses. Serageldin (1995), citing estimates by FAO, reported that on average crops effectively use only 45 percent of irrigation water. The remainder is lost during distribution and application. While less than a fifth of the municipal water in North American cities is lost to leakage, more than half is lost in Third World megacities such as Cairo, Jakarta, Lima, Manila, and Mexico City (Cohen 1995).

Third, like energy, water use is sensitive to price. Cohen (1995) cited estimates that a 10 percent increase in the price of water reduces household use by 1.5–7 percent and irrigation use by 3–20 percent. Conservation possibilities appear particularly great in industry. Despite a 30 percent increase in value-added, the U.S. manufacturing sector used a third less water in 1990 (30 billion gallons per day) than in 1980 (45 billion gallons per day) (Serageldin 1995). This has been attributed to increased charges for wastewater disposal. Total withdrawals in the United States for all purposes (irrigation, industry, domestic) were less in 1990 than in 1975 (Cohen 1995). This mirrors the de-linkage for energy shown in Figure 2.9.

Fourth, the cost of technologies for augmenting conventional freshwater supplies has dropped significantly. The cost of desalinating seawater using distillation methods fell from more than US$3.30/m³ in the 1950s to about US$1.65/m³ in the mid-1980s, while the cost of reverse osmosis fell from US$2.20–3.60/m³ in the early 1970s to US$1.10–1.65/m³ in the mid-1980s (Office of Technology Assessment 1988). Costs were much lower for desalinating brackish water, only about US$0.55/m³ in the mid-1980s. This is comparable to the upper range of costs for treated water from conventional sources.

Desalination costs depend heavily on the cost of energy. Thus, in the oil-rich Middle East, costs in 1984 were just US$1.07–3.00/m³ for seawater and US$0.27–2.13/m³ for brackish water (Gleick 1993c, Table H.35). The continued economic abundance of fossil fuels and the declining costs of alternative energy sources[17] imply that the economic feasibility of desalination will not worsen and might in fact improve even if desalination technology does not advance.[18] Costs of desalination are in fact lower today than in the mid-1980s, thanks to the decline in energy prices since the collapse of the oil market in 1985–86. Gleick (1993c, Table H.26) cited a range of US$1.30–8.00/m³ for seawater and US$0.25–1.00/m³ for brackish water in developing countries in the early 1990s (at 1985 price levels). Costs of some other unconventional supply systems, such as transport tankers, are comparable to these. The cost of water reuse (recycling of treated wastewater), which could extend supplies virtually infinitely, is only in the order of US$0.07–1.80/m³, though it might not be an appropriate source for drinking water.

Institutional failures

Given the abundant evidence that environmental degradation is more severe in developing countries, it is not surprising that more than a third of respondents to public opinion surveys in developing countries state that local environmental problems such as poor air and water quality are "very serious" in their local communities, compared to less than a fifth of respondents in developed countries (Bloom 1995). Similarly, environmental policymakers in Asia list local environmental problems such as water pollution and freshwater depletion, air pollution, deforestation, solid waste, and soil erosion as the most pressing environmental issues facing their countries (Asian Development Bank 1997).

Although poverty is associated with these forms of environmental degradation, it is far from the only culprit. Institutional failures related to poorly functioning markets and misguided policies have in fact undermined sustainable development in more fundamental ways. They are common in developed as well as developing countries, but they have particularly pernicious effects in the latter, where they are often at the root of both poverty and environmental degradation. They worsen poverty by encouraging patterns of resource use that reduce the economic returns poor households earn from using forests, fisheries, land, and water. Poverty and environmental degradation are thus intertwined in more complex ways than the previous section indicated. Although economic growth is needed to combat poverty and environmental degradation, starting and sustaining growth might not be possible if institutional failures are not first overcome.

Market failures

Economists have long known that various forms of market failure are instrumental in causing environmental degradation. Markets fail when prices do not reflect the full social costs of producing and consuming a good. This creates externalities: costs that are ignored by the good's producers and consumers and are instead paid indirectly, and often imperceptibly, by the rest of society.

An extreme but common case occurs when goods do not pass through markets at all, due to incomplete property rights. For example, intact tropical forests provide a wide range of vital ecological services, including soil protection, improved water infiltration, microclimate stabilization, and wildlife habitat. These services seldom generate economic returns for their owners, however. Tropical forests are often in a *de facto*, if not *de jure*, state of open access, in which property rights do not exist or are not enforced. Where property rights exist and are enforced, they typically cover only the land the forest grows on or the timber in its trees, not the ecological services it provides. The consequence is that forest owners have no incentive to take those services into account when they harvest timber or convert the forest to other uses.

Market failure also occurs when markets exist but are defective, causing prices to deviate from true scarcity values. Such distortions send incorrect signals to private sector decision-makers (investors, farmers, mining companies, timber enterprises, etc.). Hence, decisions based on maximizing private returns wind up causing uncompensated losses for society as a whole, often in the form of overuse or waste of natural endowments.

The classic market failure affecting renewable resources is the "tragedy of the commons." Sustaining the harvest of a renewable resource, say a fish population, requires harvesting no more than its growth increment. Growth is typically a function of, among other factors, the size of the resource stock. Growth tends to be negligible both when the stock is very low and when it is very high, as the reproductive population is small in the former case and is constrained by the ecosystem's carrying capacity in the latter case. Growth reaches a maximum at some intermediate stock level. Hence, the curve relating the growth increment to the stock has an inverted-U shape. This implies that every growth increment other than the maximum one can be generated by two stock sizes: a relatively small one and a relatively large one. The larger stock is better for both ecological and economic reasons: it renders the resource less vulnerable to extinction and makes it easier to harvest.

The larger stock is thus what a rational individual or community should prefer.[19] But as is well-known from the work of Gordon (1954), Hardin (1968), and Weitzman (1976), when a resource is in a state of open access, users have an individual incentive to deplete the stock to a low level. Each tries to capture as much of the resource as possible before the others do. Depletion makes the resource more difficult to harvest, but this cost is an externality that users ignore when they lack exclusive property rights, as they then cannot (as individuals) control it. The result is that the resource is overexploited and generates less rent than it potentially could. In the extreme, rent is driven to zero: it is completely dissipated, and users earn only the opportunity cost of their labor. In a slow-growing developing country, this would be the subsistence wage.

The "tragedy of the commons" has the look and feel of a Malthusian scenario – many people are using the resource, they are poor, and the resource is depleted – but it does not result from population growth.[20] Instead, resource depletion and poverty both stem from the same institutional failure, open access. Most important, unlike a mechanistic Malthusian model, there is a way out of the tragedy: institutional reform to strengthen individual or collective property rights.[21] This is a "win-win" solution: Users will have an economic incentive to maintain a larger resource stock, thus reducing the risk of its collapse and enhancing the sustainability of harvests, and they will earn larger rents, thus raising their incomes and contributing to economic growth. Far from being in conflict, improved resource management and economic growth thus go hand-in-hand when open access problems are addressed.

Fisheries and rangelands are classic examples of resources prone to open-access rent dissipation. The sorry state of the world's fisheries, and well-known cases of overgrazing in the Sahel and the Indian subcontinent, indicate how widespread and how severe the open access problem is. Forests and groundwater resources commonly suffer from it, too. To our knowledge, no one has attempted to quantify how much of the aggregate natural resource base in developing countries is in a state of near or complete open access, but the amount, and the toll in terms of both ecological destruction and human misery, are surely enormous. So too, therefore, would be the economic benefits of addressing this problem.

Policy failures

Market failures, whether due to monopoly, open access, free riders, or high transaction costs, are pervasive and fundamental, but they involve few mysteries. Economists have studied them for decades. In contrast, the role of government, or policy, failures in causing environmental degradation has become widely appreciated only relatively recently. In fact, many market failures are best viewed as manifestations of policy failures. For example, open access depletion of forests in Nepal is rooted in governmental decisions to nationalize resources and delegitimize traditional systems of user rights (Wallace 1983). Decisions by both colonial and post-colonial governments have also undermined traditional property relations in Africa (see Peters 1994 on land and water in Botswana, and Gillis 1988b on forests in Ghana).

One of the prime causes of policy failures leading to needless ecological and economic damage in poor countries is the tendency of policymakers to overlook the environmental consequences of tax, trade, and other *non*environmental policies. Policies intended to attain nonenvironmental goals can have very large, and generally unanticipated, impacts on the environment (Reed 1992, Munasinghe and Cruz 1994). In much of Africa, Latin America, and Asia, the pursuit of agricultural, industrial, and urbanization objectives has had significant corrosive effects on agricultural soils, forests, water and air quality, coastal fisheries, and coral reefs. Measures that reduce environmental damage from such policies are, like policies that overcome poverty, both good ecology and good economics. This is especially obvious in the former communist countries of Central and Eastern Europe and the former Soviet Union, where the policies of central planners created economic as well as ecological disasters (Åhlander 1994, Feshbach 1995, Libert 1995).

A second, and closely related, source of policy failures is the persistent lack of understanding by policymakers of the role of prices in resource conservation and environmental protection. Shortsighted subsidy programs that deeply underprice forests, water, energy, and other resources comprise a high proportion of policy failures (OECD 1996, de Moor 1997, de Moor and Calamai 1997). The tendency to underprice natural resources might in part be a hangover from the days, not too many years ago, when economics textbooks habitually presented air and water as prototypical "free goods," available in limitless supplies at zero prices. Today we know that *clean* air and *clean* water are scarce resources, not free goods. Even so, governments persist in underestimating the role of prices in resource conservation and resource allocation more generally.

Underpricing of forests

Subsidies and other policy failures in forestry have been especially destructive to environmental and economic goals in dozens of tropical countries. Brazil's government long provided heavy subsidies to ranching and other activities encroaching on the Amazon rainforest (Browder 1988, Mahar 1991, Binswanger 1991). Three to four thousand square miles of the Amazon were deforested each year throughout the 1970s. When ranching replaced the rainforest, it destroyed more forest-based jobs than it created. Nevertheless, the government made the conversion of forests to ranches as cheap as possible. It offered ranchers 15-year tax holidays, investment tax credits, exemptions from export taxes and import duties, and loans with below-market interest rates. Although a typical subsidized ranching investment yielded a net loss to the economy equivalent to 55 percent of the initial investment, thanks to the heavy subsidies ranchers earned private returns of up to 250 percent of their investment.

Many governments have also subsidized timber production. Most tropical countries, including Indonesia, the Philippines, some states in Malaysia, Surinam, Guyana, and Ghana and other African nations, have charged very low fees for timber in public forests (Page *et al.* 1976, Ruzicka 1979, Gillis 1980, 1988a, 1988b, 1993, Vincent 1990, Sizer and Rice 1995, Sizer 1996). They typically levy the fees on extracted logs rather than on standing timber volumes (stumpage), thus reducing the incentive for concessionaires to minimize logging damage. The low fees contribute to the chronic underfunding of forestry departments responsible for managing and protecting public forests. Tropical countries also usually limit concession agreements to short time periods, without offering transparent conditions for renewing the agreements. This undermines concessionaires' interest in future timber harvests and thus in sustainable forest management. Thailand's forestry policies were so wanton that its rainforest has all but disappeared. The same can be said for the Ivory Coast and the Philippines; Gabon is on the same path.

Trade policy has often been destructive of economic and environmental values in tropical forests. In the 1980s, major tropical timber exporters such as Indonesia and Peninsular Malaysia banned log exports as a means of promoting domestic wood-processing industries (Gillis 1988a, Vincent 1992). The bans created implicit input subsidies for local producers by driving down the domestic price of logs. These subsidies made the processing of logs into semi-finished exportable products immensely profitable for local sawmills and plywood mills. But they also encouraged grossly inefficient processing, resulting in the waste of huge quantities of timber. By reducing stumpage values (log price minus logging costs), they also reduced the scope for collecting adequate revenue to cover forest management expenses.

Underpricing of water

Underpricing of water resources has long been common throughout the world, with users generally paying less than 10 percent of operating costs and even less of total costs including capital costs (World Bank 1992). Indeed, it is safe to say that where one finds an acute crisis in water availability, subsidies are usually the prime suspect. Subsidies are especially heavy for irrigation water, although most countries also subsidize municipal water. Over half of all investment in agriculture in developing countries in the 1980s was in water resource development. Public irrigation systems operated by government agencies and state-owned enterprises in developing countries had absorbed US$250 billion in public funds by the mid-1980s (Gillis 1991a). Governments spent perhaps another US$15 billion a year on such projects in the late 1980s.

Yet, in a sample of World Bank irrigation projects, revenues covered only 7 percent of project costs on average (Gillis 1991a). Irrigation policies have deeply underpriced water resources in Mexico for decades (Ascher 1997). Eighty percent of all public investment in agriculture in Mexico from 1940 to 1990 was in irrigation projects. During 1945–59, farmers paid only 45 percent of the operating costs of irrigation systems. This figure slipped to 30 percent by the early 1960s. After rising to 70 percent by 1969–71, it fell in the 1980s to less than 20 percent.

Not surprisingly, cheap prices for irrigation water have induced extremely wasteful use. The Aral Sea provides the most dramatic example (Libert 1995). The sea, which was once the world's fourth largest inland body of water, is now three separate, shallow pools. Soviet planners diverted more than 80 percent of the water in the two principal rivers flowing into the sea to cotton fields in Kazakhstan and Uzbekistan. They charged communal farms an effective price of zero rubles for the water. These policies virtually eliminated inflows to the sea, thus destroying the sea's fishing and transportation industries. They also led to high rates of respiratory disorders from the inhalation of pesticide residues deposited on the former seabed. Another stark example is the Indus basin, where low water prices encouraged excessive irrigation and created waterlogging and salination problems in millions of hectares as early as the 1950s (Khan 1991).

In economic terms, overuse of water occurs because farmers, who have a private incentive to use water up to the point where its marginal value as a production input equals its price, face a subsidized price far below the social cost of providing irrigation water. The flip side of this is that society gains a lot, and farmers lose little, if water is priced and allocated more efficiently. In China, a cubic meter of water is worth ten times more in municipal and industrial uses than in agriculture (World Bank 1992). In northern Thailand, the *gross* marginal value of irrigation water during 1980–91 was on the order of 1–1.5 baht per cubic meter, while the *net* marginal value of water in municipal and industrial uses was on the order of 4–7 baht per cubic meter (Vincent *et al.* 1995).

Underpricing is also common in developing countries in the case of urban drinking water. Subsidies on drinking water are usually enacted for distributional reasons: to support water consumption by the very poor. Unfortunately, perverse distributional results often occur, with more affluent neighborhoods disproportionately receiving improved water services (Gillis 1991a).

Underpricing of energy

Energy pricing provides another frightful history of policy failure leading to economic as well as environmental damage. In oil-rich countries such as Nigeria and Venezuela, domestic use of energy has been kept cheap as a stimulus to industrialization and economic diversification. This has had multiple adverse effects. First, it has encouraged wasteful domestic consumption, thereby reducing petroleum and gas reserves and potential export earnings. Second, it has artificially promoted the use of motor bikes and automobiles, adding to urban congestion. Third, it has promoted industries that are ill suited to the countries' labor and capital endowments. Finally, because the combustion of fossil fuels is the primary source of air pollution emissions, it has contributed to environmental degradation.

Indonesia's kerosene policy furnishes an instructive example (Gillis 1991a). From 1970 to 1984, the Indonesian government heavily subsidized the consumption of kerosene and other fuels. It justified the subsidies as a way of aiding poor rural dwellers, who it thought used kerosene for cooking, and reducing environmental degradation, as it expected lower kerosene prices to discourage fuelwood cutting. These assumptions turned out to be wrong. Rural families used kerosene primarily for lighting, not for cooking. Only 20 000 hectares of forestland were "saved" each year by the subsidy, at an implicit annual cost of almost US$500 000 per hectare. Replanting programs, in contrast, cost only US$2500 per hectare. Moreover, 80 percent of kerosene turned out to be consumed by the relatively wealthy, not the poor. The low price of kerosene forced the government to subsidize diesel fuel as well, because the two fuels can be substituted in truck engines. This amplified the environmental damage caused by the subsidy policy. These multiple costs finally led the government to reduce the subsidies sharply. Indeed, Indonesia now tries to price most fuels at world market levels.

Even some oil importers, such as Argentina, China, and India, have underpriced petroleum products by as much as 50 percent of the world price, encouraging imports they cannot afford, nurturing industries that cannot compete in world markets, and causing environmental degradation that market pricing would have discouraged (Gillis 1991a). As shown earlier in Figure 2.8, the energy/GDP ratio varies considerably across countries. Cross-country studies have identified mistaken energy policies for the unusually high ratios in countries such as Argentina, India, South Africa, Venezuela, and Zambia, where prices were greatly distorted until the wave of economic reforms in the 1980s and 1990s (Gillis 1991a). As late as 1996, Venezuela was still pricing gasoline at about 25 U.S. cents a gallon.

The consequences of energy underpricing were particularly evident in the command economies in Central and Eastern Europe and the former Soviet Union. Coal prices, for example, were less than 10 percent of production costs in the former Soviet Union (World Bank 1992). The collapse of the Soviet empire revealed that such low prices had encouraged the development of excessively energy-intensive industries, which were grossly uncompetitive at world energy prices, and had generated levels of air and water pollution that were among the highest in the world.

Underpricing agrochemicals

Agricultural subsidies have generated similar economic and environmental damage. Governments around the globe have adopted policies that severely underprice agrochemicals, especially fertilizers made from natural gas. As in the cases of water and energy, fertilizer subsidies have led to substantial waste. In Indonesia, for example, fertilizer use increased by 77 percent during 1980–85, when the government offered generous subsidies. As a result, Indonesian rice farmers applied three times as much fertilizer per hectare as rice farmers in Thailand and Philippines, where subsidies were smaller (Gillis 1991a).

Governments usually justify agrochemical subsidies on the grounds of output effects, and sometimes too on the grounds that they serve soil enrichment and conservation purposes. These arguments rarely withstand scrutiny. Although the subsidies might indeed generate gross benefits in terms of higher crop yields, those benefits are typically less than the subsidy-inclusive sum of the direct and indirect costs of producing the incremental output. The yield increases can also be short-lived. Many agrochemical subsidies have been not only expensive but also strongly, and strangely, counterproductive. This was the case with pesticide subsidies in Indonesia, where several heavily subsidized pesticides that rice farmers routinely used in the 1980s were being used by researchers in other countries to *increase* the number of pests in experimental plots, due to their greater effects on the pests' natural predators. There is similar evidence that sustained use of chemical fertilizers can reduce soil fertility (Gillis 1991a). The conservation argument for fertilizer subsidies is particularly weak in semi-arid tropical countries, where organic fertilizers and moisture-retaining farming methods are more appropriate but rarely subsidized.

National benefits and costs of sustainable development

Bad economic development policies, which have exacerbated poverty, and institutional failures, which have made the costs of environmental degradation appear lower than they truly are, form the principal explanation for unsustainable development in developing countries. As these are essentially self-inflicted problems, which can be reversed by implementing improved policies,[22] developing countries can in principle make great strides toward sustainable development on their own, without external assistance.

They stand to gain substantial economic benefits from doing so. The Asian Development Bank (1997, Table 4.2) has compiled estimates of the economic benefits from abating urban air and water pollution in several Asian countries. The median estimate is on the order of 2 percent of GDP. If the benefits of pollution abatement are of comparable magnitude in other developing regions, then the benefits potentially available for the entire developing world exceed US$100 billion (based on 1995 GDP estimates reported by the World Bank 1997b). Adding the benefits of other types of environmental improvements would of course only increase this figure.

If the benefits of improved environmental management are so great, why aren't developing countries pursuing them more aggressively? To be fair, some developing countries have indeed achieved success, some of it striking, in addressing their environmental problems. For example, Malaysia acted decisively in the 1970s to stem water pollution from crude palm oil mills, a major export industry (Vincent *et al.* 1997b, Chapter 10). It also introduced a fisheries licensing policy that, in combination with a fast-growing economy that provided superior employment opportunities for poor people in coastal villages, greatly reduced overfishing in inshore waters (*ibid.*, Chapter 6). And, as discussed earlier, a decade of

rapid economic growth and diversification, combined with the establishment of a permanent forest estate and secure property rights for agriculture, has contributed greatly toward stabilizing forest areas in Peninsular
Malaysia (*ibid.*, Chapters 4 and 5).

Yet, the pollution data in Figures 2.2–2.7 indicate that most developing countries, especially the poorest ones, have a long way to go. Political
leaders of developing countries typically offer two reasons for not acting
more decisively to address environment degradation. One may be termed
the competitiveness argument; the other, the financing gap.

The competitiveness argument

According to the competitiveness argument, in a global economy environmental protection undermines economic growth by raising production costs, which makes a country less attractive to foreign investors and
makes its exports less competitive in international markets. Hence, the
argument goes, developing countries that pursue sustainable development end up losing out in terms of economic growth. If this reasoning is
correct, then transfers of financial resources from North to South would
indeed appear to be necessary to enable developing countries to achieve
both higher incomes and a better quality environment.

Although this argument superficially has a certain logic, it ignores the
economic benefits of environmental protection. It overlooks the fact that
domestic winners (the public, largely silent) outweigh domestic losers
(polluting industries, typically more vocal) when nations close the gap
between the social and private costs of environmental degradation. Moreover, it exaggerates the impact of environmental regulations on investment decisions and trade patterns. In a recent World Economic Forum
(WEF) survey of nearly 3000 executives at domestic and multinational
companies in 53 countries (28 high-income, 25 low-income), environmental regulations ranked twenty-third out of twenty-seven factors influencing foreign direct investment (Panayotou and Vincent 1997). Executives
rated profit repatriation, contract enforcement, domestic market size and
growth, workforce productivity and skill, corporate taxation, risk of
expropriation, and infrastructure as much more important. Wheeler and
Mody (1992) and Repetto (1995) reported similar results.

Regarding trade, Kalt (1988) and Tobey (1990) failed to detect a statistically significant relationship between exports and environmental regulations. After conducting a comprehensive survey of the literature, Jaffe *et
al.* (1995) concluded that trade patterns are better explained by trade and

investment policies, along with normal structural changes that occur during economic development, than by environmental regulations. Executives in the WEF survey reported that environmental regulations have only a modest impact on their companies' profitability, which implies a similarly modest impact on export competitiveness. Three-fifths of the executives stated that environmental regulations are neutral with regard to profitability. Thirteen percent claimed that environmental regulations actually *enhance* their companies' profitability.

An econometric analysis of the executives' responses revealed that the impact of environmental regulations on profitability depends critically on the regulations' characteristics (Panayotou and Vincent 1997). Flexibility – whether companies can select the least expensive means of attaining environmental objectives – and transparency and stability – whether companies can readily identify those objectives in the environmental regulations they face – were the most important characteristics. Other evidence confirms the importance of flexibility. For example, one study has estimated that pollution abatement costs in the Indian power sector could be reduced by two-thirds if that country replaced its "command-and-control" regulations, which mandate uniform emissions standards and "best available" technology, with more flexible, market-based instruments (Asian Development Bank 1997). The same study estimated even more dramatic potential cost savings in China, 90 percent.

The history of trade reform suggests that such evidence, however compelling, will not make overnight converts of politicians convinced that environmental regulations harm international competitiveness. It also suggests, more positively, a way to make progress toward sustainable development despite politicians' concerns. The benefits of free trade to both individual nations and the world as a whole are surely more widely accepted than the benefits of environmental protection. Yet, trade liberalization has been a gradual, contentious process spanning half a century and counting. As with environmental protection, countries fear economic losses if they reduce their trade barriers unilaterally. The various GATT rounds overcame such fears by being multilateral processes. This coordinated approach boosted countries' confidence that they would benefit from trade reform. Over time, it has encouraged them to make impressively deep cuts in tariffs and other trade barriers.

Soon after the Earth Summit, the United Nations Department for Policy Coordination and Sustainable Development established an expert group to advise the newly established Commission for Sustainable

Development (CSD) on financial aspects of *Agenda 21*. One of the group's first suggestions was to address the competitiveness argument by establishing a multilateral process for coordinating domestic environmental policy reforms. Like GATT, the proposed process would be voluntary and would be founded on the principles of national sovereignty, nondiscrimination, and reciprocity. Depending on its size and scope, it could be administered by either an existing multilateral organization or a new secretariat. As yet, the CSD has not endorsed this idea.[23]

The financing gap

The financing gap is the issue raised at the very beginning of this chapter: that developing countries require substantial, and concessional, transfers of financial resources from developed countries to get on a sustainable development track. *Agenda 21* placed the amount at US$125 billion per year. Since Rio, virtually no progress has been made toward achieving this goal. In 1995, official development assistance (ODA) from bilateral and multilateral donor organizations was only US$60.1 billion (Killick 1997). This was barely higher than the amount in 1992, US$58.9 billion. In real (inflation-adjusted) terms, it was lower.

This situation would appear to represent a retreat from sustainable development. Fortunately, it is less bleak than it appears. The United Nations did not derive the US$125 billion figure by analyzing the gap between the total amount of investment required for sustainable development and the ability of developing countries to finance that investment from domestic sources. In formulating *Agenda 21*, it did indeed estimate the total amount of required investment, arriving at the breathtaking figure of US$625 billion per year (the "budget" for *Agenda 21*). It assigned US$125 billion to concessional transfers simply by applying its long-standing target that developed countries should contribute 0.7 percent of their GDP to ODA. It did not assess the availability of financial resources within developing countries. The US$125 billion figure should therefore not be interpreted as required gap financing. It is simply the aggregate value of the 0.7 percent target, at GDP levels in rich countries in the early 1990s.

Furthermore, the US$625 billion figure, and thus the US$125 billion figure mechanically derived from it, include at least two significant upward biases. First, although *Agenda 21* highlights the importance of environmental and nonenvironmental policy reforms, its cost estimates are by and large based on the costs of environmental protection under existing policies – that is, in the presence of existing market and policy failures. Resolving those failures would reduce environmental degradation in the first place and thus reduce the need for costly remedial actions. Second, the cost estimates in *Agenda 21* are by and large based on costs associated with command-and-control regulations. As the estimates of pollution abatement costs in the Indian and Chinese power sectors indicate, this is far from a least-cost approach. For both reasons, *Agenda 21*'s price tag for sustainable development is too high.

While developing countries would arguably be better off with concessional transfers than without them (although one must consider the ill effects of aid dependency; see Killick 1997), the pertinent issue is what developing countries should do if concessional transfers continue to fall short of the *Agenda 21* targets. Should they hope, surely quixotically, that US$125 billion in annual transfers will someday materialize? Should they forgo the investments that *Agenda 21* intended the transfers to finance? Or should they finance, however reluctantly, some portion of those investments themselves? Given the *prima facie* evidence that developing countries will benefit economically from reducing environmental externalities, and the empirical evidence that such externalities are large in value terms, the third option is probably the best. Developing countries will only harm themselves if they refuse to invest in sustainable development until developed countries foot the bill.

Of course, not all countries are in an equally advantageous position to mobilize domestic financial resources. Three economists from the International Monetary Fund (IMF), writing in their personal capacity (not as IMF staff members), have presented rough estimates of the *incremental* financial resources that can be mobilized for sustainable development by various domestic measures (Gandhi *et al.* 1997). The estimates, which are reproduced in Table 2.3, are for developing and developed countries combined. The quantifiable total is US$1.2 trillion per year, with unknown but "potentially quite large" amounts available from macroeconomic and structural reforms, and additional but unknown amounts available from environmental taxes other than ones on petroleum products and carbon. Note the major contribution by measures aimed at reducing the underpricing of the natural environment – levying user fees and charges, imposing environmental taxes, and eliminating and improving the targeting of subsidies – which together account for more than half the total.

The quantifiable total in Table 2.3 is nearly twice the US$625 billion budget for *Agenda 21*. But the "best guess" by Gandhi *et al.* (1997) is that

Table 2.3. *"Order-of-magnitude" estimates of domestic financial resources available for sustainable development in developed and developing countries combined, according to Gandhi et al. (1997)*

Measure	Magnitude
Public revenue mobilization	
Reforming tax systems	US$300 billion
Levying user fees and charges	US$150 billion
Imposing environmental taxes	
Petroleum excise taxes	US$100 billion
Carbon taxes	US$55 billion
Others (e.g., pollution taxes)	>0
Public expenditure savings	
Reducing unproductive expenditures	US$90 billion
Eliminating and better targeting subsidies	US$435 billion
Curtailing military expenditures	US$70 billion
Other reforms	
Macroeconomic reforms	>0
Structural reforms	>0
Total	>US$1.2 trillion

only about US$250 billion of the US$1.2 trillion can be mobilized within developing countries. Even if we accept that the US$625 billion figure is biased upward by some large but unknown amount, and bear in mind that the US$250 billion figure excludes several potentially important sources of financing, a gap of many billion dollars could well exist.

Are concessional transfers the only way to plug this gap? Not for all countries. International private capital flows, which grew from less than US$10 billion in 1970 to more than US$150 billion in 1995 (Killick 1997), offer developing countries a third source of financial resources in addition to domestic resources and concessional transfers. Private flows now exceed ODA by a factor of more than two. They are a significant source of capital only for a relatively small number of middle-income countries, however.[24] They have been but a trickle in developing countries that are already heavily indebted or have poor credit ratings. In 1993, low-income countries received only 8 percent of the net flow of long-term international private capital to developing countries (Griffin and McKinley 1996, cited in Killick 1997). It is in these countries, not the entire developing world, that concessional transfers are needed to ensure that viable investments in sustainable development do not go unfunded.

Global benefits of domestic action

Concessional transfers can be justified on economic efficiency grounds when environmental externalities cross national boundaries. In that context, concessional transfers are best viewed not as aid but as payments for services rendered: payments for the benefits developed countries enjoy as a consequence of environmental improvements made by developing countries. In the absence of such payments, a developing country lacks an economic incentive to protect its environment beyond the point that maximizes its own net benefits. It then has only an economic incentive to consider externalities affecting its own population, not externalities affecting populations in other countries. Concessional transfers provide an economic incentive to take transboundary externalities into account and thereby move toward the globally, as opposed to the nationally, optimal level of environmental protection. They compensate developing countries for the incremental costs incurred in protecting the environment beyond the nationally optimal point.

Large reductions in transboundary externalities do not necessarily require large amounts of international compensation, however. The amount of international compensation might in fact be relatively small. Consider the case of tropical forests. The protection of tropical forests benefits both developing countries and the rest of the world, but it also imposes costs on developing countries. The most important cost is usually the opportunity cost of forestland: the net benefits generated by alternative uses of the land. Plotting this opportunity cost against the cumulative area of forestland (from least to most valuable in alternative uses) maps out the supply curve for forestland. The upward-sloping line in Figure 2.10 represents such a curve for a hypothetical developing country.

Most tropical forests are public, not private, lands (Repetto and Gillis 1988, Gillis 1993). Government agencies usually take timber values into account in determining how much forest to retain on a permanent or semi-permanent basis. In Indonesia, Malaysia, and many other developing countries, forests that are retained because the timber they provide is thought to be more valuable than other products the land could potentially produce (e.g., food) are typically called "production" forests. In

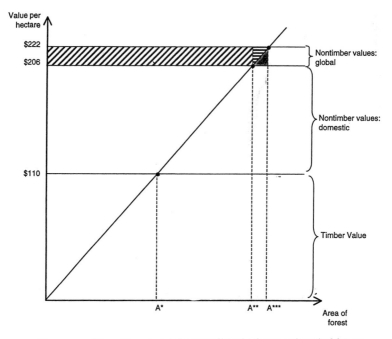

Figure 2.10 Domestic and global externalities in the case of tropical forests.
Note: Values based on data in Lampietti and Dixon.

Figure 2.10, A^* represents the optimal area of forestland when timber is
the only significant forest value. The per-hectare value for timber production shown in the figure, US$110, is the median annualized value reported
by Lampietti and Dixon (1995) in their survey of forest valuation studies.
Forests should be retained on lands where the value in alternative uses is
less than this amount.

Many of the nontimber goods and services provided by tropical forests
benefit mainly the developing countries where the forests occur. Examples include "minor" forest products (bamboo, fuelwood, medicinal
plants, etc.), hunting and fishing, recreation, and watershed services.
According to Lampietti and Dixon, the median annualized value of such
nontimber goods and services is on the order of US$96 per hectare.
Adding this value to the median timber value implies that forests should
be retained on lands with alternative uses worth less than US$206 per
hectare. This is the nationally optimal level of forest protection, denoted
by A^{**}.

Other nontimber goods and services might accrue primarily to people in the rest of the world. Option and existence values related to biodiversity are a prime example: People in richer countries might value genetic resources as a potential source of new pharmaceutical products and endangered species for altruistic reasons. Lampietti and Dixon report a median annualized value of US$16 per hectare for such option and existence values. With these values taken into account, the globally optimal forest area is A^{***}.

In the absence of compensation from other countries, tropical countries have no economic incentive to retain the incremental forest area between A^{**} and A^{***}. The opportunity cost of protection, given by the forestland supply curve, exceeds the national benefits of protection, US$206 per hectare, by the solidly shaded triangle under the supply curve. Inducing tropical countries to achieve the globally optimal forest area requires compensation at least as large as this triangle.

International compensation for the incremental costs of globally optimal environmental protection is the principle on which the Global Environmental Facility is founded and operated. This principle is conceptually sound. As Figure 2.10 illustrates, however, international compensation is not necessarily the best way to protect globally valuable forest benefits. *Nationally* optimal forest protection up to A^{**} *incidentally* provides a great amount of global benefits, the rectangle shaded with slanted stripes. This amount is much greater than the amount "purchased" by international compensation, the sum of the two shaded triangles. This result is not a quirk of the way the supply curve in Figure 2.10 is drawn. Rather, it reflects the small size of the global externality relative to the national benefits.

Some rough estimates compiled by Prof. William Hogan of Harvard University's Kennedy School of Government suggest that global externalities are also relatively small compared to national externalities in the case of fossil fuel combustion (Hogan 1996). Hogan estimated the total external costs of gasoline use in the U.S. at nearly US$10 per million BTUs. Of this total, global externalities from CO_2 emissions (related to climate change) accounted for only about US$0.25. National externalities associated with health damage from other air pollutants, highway depreciation, oil security, and so on accounted for the remainder. In the case of diesel fuel, the global externality share was even smaller; in the case of coal, heating oil, and jet fuel, it was larger but still far below half. Crapanzano and Del Furia (1997) reported similar findings for Italy. They estimated that fossil fuels burned in electricity generation caused local air pollution damages of 21 billion lira per year, compared to global warming damages of only 4.2–9.9 billion lira per year.

The relative insignificance of global externalities is also implied by comparing the global benefits (to both developed and developing countries) of abating greenhouse gas emissions, which Nordhaus (1994) has estimated at US$13 billion per year,[24] to the benefits of pollution abatement within developing countries, which we roughly estimated earlier at more than US$100 billion per year.

These examples indicate that the pursuit of environmental protection for purely national reasons is not only good for developing countries but can also be good for the global commons. It might not achieve the globally optimal level of protection, but it might come close.

Summary

To recap: the global perspective on sustainable development, which emphasizes environmental problems in the global commons, appears to be inconsistent with the economically rational environmental priorities of developing countries. The major costs of environmental degradation in developing countries occur within their national borders. By taking steps to reduce those costs, developing countries will incidentally reduce a major portion of the externalities that do spill over their borders.

The key step is to address market failures and policy failures. Developing countries can take this step without destroying the competitiveness of their economies. By addressing market failures and policy failures, developing countries can reduce the need for direct financing of environmental protection and can generate a good portion of the financial resources required to meet the needs that do remain. The need for international agreements and international compensation to achieve globally sustainable development is thus not as great as the Earth Summit implied.

REFERENCES

Åhlander, A.-M. S. 1994. *Environmental Problems in the Shortage Economy*. Edward Elgar, Aldershot, United Kingdom.

Anonymous. 1997. "At last, the fuel cell." *The Economist* (October 25):89–92.

Arrow, K. and Kurz, M. 1970. *Public Investment, the Rate of Return, and Optimal Fiscal Policy*. Johns Hopkins University Press, Baltimore, Maryland.

Arrow, K., Bolin, B. and Costanza, R. 1995. "Economic growth, carrying capacity, and the environment." *Science* 268:520–21.

Ascher, W. 1997. *Why Governments Waste Natural Resources: Political Economy of Natural*

Resource Policy Failures in Developing Countries. Center for International Development Research, Duke University, Durham, North Carolina.

Asian Development Bank. 1997. *Emerging Asia: Changes and Challenges.* Manila, Philippines.

Binswanger, H. P. 1991. "Brazilian policies that encourage deforestation in the Amazon." *World Development* 19(7):821–829.

Bloom, D. 1995. "International public opinion on the environment." *Science* 269:354–358.

Browder, J.O. 1988. "Public policy and deforestation in the Brazilian Amazon." In R. Repetto and M. Gillis, eds., *Public Policies and the Misuse of Forest Resources,* pp. 247–297. Cambridge University Press, New York.

Brown, L. and Wolf, E. 1984. *Soil Erosion: Quiet Crisis in the World Economy.* Worldwatch Paper No. 60. Worldwatch Institute, Washington, DC.

Cohen, J. 1995. *How Many People Can the Earth Support?* W.W. Norton & Company, New York.

Crapanzano, G. and Del Furia, L. 1997. "The ExternE approach to the assessment of energy external costs." *FEEM Newsletter* 1997 (3):13–16.

Crosson, P. 1995. "Future supplies of land and water for world agriculture." In Nurul Islam, ed., *Population and Food in the Early Twenty-First Century,* pp. 143–159. International Food Policy Research Institute, Washington, DC.

Crosson, P. 1997. "Will erosion threaten agricultural productivity?" *Environment* 39(8):4–9, 29–31.

Daly, H. 1991. "Towards an environmental macroeconomics." *Land Economics* 67(2):255–259.

Dasgupta, P. 1995. "The population problem: theory and evidence." *Journal of Economic Literature* 33(4):1879–1902.

Dasgupta, P. and Mäler, K.-G. 1991. "The environment and emerging development issues." In *Annual World Bank Conference on Development Economics 1990.* The World Bank, Washington, DC.

de Moor, A. 1997. "Key issues in subsidy policies and strategies for reform." In J. Holst, P. Koudal and J. R. Vincent, eds., *Finance for Sustainable Development: The Road Ahead,* pp. 285–313. United Nations, New York.

de Moor, A. and Calamai, P. 1997. *Subsidizing Unsustainable Development: Undermining the Earth with Public Funds.* The Earth Council, San José, Costa Rica.

Dregne, H. and Chou, N. T. 1992. "Global desertification dimensions and costs." In H. Dregne, ed., *Degradation and Restoration of Arid Lands.* Texas Tech University Press, Lubbock, Texas.

Ehrlich, P. R., Daily, G. C., Daily, S. C., Myers, N. and Salzman, J. 1997. "No middle way on the environment." *The Atlantic Monthly* (December):98–104.

Energy Information Administration. 1997. *International Energy Outlook 1997.* www.eia.doe.gov/oiaf/ieo97.

FAO. 1989. *Yearbook of Forest Products 1987.* Rome.
    1997. *State of the World's Forests 1997.* Rome.

FAO/UNEP. 1981. "Tropical forest resources assessment project: forest resources of tropical Asia." *UN 32/6.1301–78–04, Technical Report No. 3.* Rome.

Feshbach, M. 1995. *Ecological Disaster: Cleaning Up the Hidden Legacy of the Soviet Regime.* The Twentieth Century Fund Press, New York.

Gallup, J., Radelet, S. and Warner, A. 1997. "Economic growth and the income of the poor." Manuscript. Harvard Institute for International Development, Cambridge, Massachusetts.

Gandhi, V.P., Gray, D. and McMorran, R. 1997. "A comprehensive approach to domestic resource mobilization for sustainable development." In J. Holst, P. Koudal and J. R. Vincent, eds., *Finance for Sustainable Development: The Road Ahead*, pp. 169–219. United Nations, New York.

Gillis, M. 1980. "Fiscal and financial issues in tropical hardwood concessions." *Development Discussion Paper* No. 110. Harvard Institute for International Development, Cambridge, Massachusetts.

  1988a. "Indonesia: public policies, resource management, and the tropical forest." In R. Repetto and M. Gillis, eds., *Public Policy and the Misuse of Forest Resources*, pp. 43–114. Cambridge University Press, Cambridge, United Kingdom.

  1988b. "West Africa: resource management policies and the tropical forest." In R. Repetto and M. Gillis, eds., *Public Policy and the Misuse of Forest Resources*, pp. 115–204. Cambridge University Press, Cambridge, United Kingdom.

  1991a. "Tacit taxes and sub-rosa subsidies." In R. Bird, ed., *More Taxing than Taxes: The Taxlike Effects of Nontax Policies in LDCs*. Institute for Contemporary Studies, San Francisco, California.

  1991b. "Tropical deforestation: economic, ecological and ethical dimensions." *South Atlantic Quarterly* 90(1):13–14.

  1993. "Forest incentive policies." In N. Sharma, ed., *Managing the World's Forests*, pp. 139–175. Kendall-Hunt Publishing Co., Dubuque, Iowa.

  1997. "Sustainable development in poor and formerly poor nations." *Houston Business Review* 6:40.

  1998. "Turbulence in emerging markets: East Asian antecedents." Paper presented at the Federal Reserve Bank of Dallas, Texas, January 30, 1998.

Gleick, P.H., ed. 1993a. "An introduction to global fresh water issues." In *Water in Crisis: A Guide to the World's Fresh Water Resources*, pp. 3–12. Oxford University Press, New York.

  1993b. "Water and energy." In *Water in Crisis: A Guide to the World's Fresh Water Resources*, pp. 67–79. Oxford University Press, New York.

  1993c. *Water in Crisis: A Guide to the World's Fresh Water Resources*. Oxford University Press, New York.

Gordon, H.S. 1954. "The economic theory of a common-property resource: the fishery." *Journal of Political Economy* 62:124–142.

Griffin, K. and McKinley, T. 1996. "New approaches to development cooperation." *UNDP Discussion Paper No. 7*. United Nations, New York.

Grossman, G. and Krueger, A. 1991. "Environmental impacts of a North American Free Trade Agreement." *NBER Working Paper No. 3914*. National Bureau of Economic Research, Cambridge, Massachusetts.

  1995. "Economic growth and the environment." *Quarterly Journal of Economics* 110:353–77.

Hardin, G. 1968. "The tragedy of the commons." *Science* 162:1243–1248.

Hogan, W. 1996. "Energy economics and energy policy." Lecture notes, June 24, 1996. Harvard Institute for International Development, Environmental Economics and Policy Analysis Workshop, Cambridge, Massachusetts.

Huang, J. Rozelle, S. and Rosegrant, M. 1997. "China's food economy to the twenty-first century: supply, demand, and trade." *Food, Agriculture, and the Environment*

Discussion Paper No. 19. International Food Policy Research Institute, Washington, DC.

Jaffe, A., Peterson, S., Portney, P. and Stavins, R. 1995. "Environmental regulation and the competitiveness of U.S. manufacturing: what does the evidence tell us?" *Journal of Economic Literature* 33:132–163.

Kalt, J. 1988. "The impact of domestic environmental regulatory policies on U.S. international competitiveness." In A.M. Spence and H. Hazard, eds., *International Competitiveness*, pp. 221–262. Ballinger, Cambridge, Massachusetts.

Khan, F.K. 1991. *A Geography of Pakistan.* Oxford University Press, Karachi, Pakistan.

Killick, T. 1997. "What future for aid?" In J. Holst, P. Koudal and J.R. Vincent, eds., *Finance for Sustainable Development: The Road Ahead*, pp. 77-107. United Nations, New York.

Lampietti, J.A. and Dixon, J.A. 1995. "To see the forest for the trees: a guide to non-timber forest benefits." *Environment Department Paper No. 013.* The World Bank, Washington, DC.

Larson, B. and Frisvold, G. 1996. "Fertilizers to support agricultural development in sub-Saharan Africa: what is needed and why." *Food Policy* 21(6):509–525.

Libert, B. 1995. *The Environmental Heritage of Soviet Agriculture.* CAB International, Wallingford, Oxford, United Kingdom.

Lindert, P.H. 1996a. "Soil degradation and agricultural change in two developing countries." *Working Paper No. 82.* Agricultural History Center, University of California, Davis.

1996b. "The bad earth? China's agricultural soils since the 1930s." *Working Paper No. 83.* Agricultural History Center, University of California, Davis.

Mahar, D.J. 1991. *Government Policies and Deforestation in Brazil's Amazon Region.* World Bank, Washington, DC.

Meadows, D., Meadows, D., Randers, J. and Behrens III, W. 1972. *The Limits to Growth.* Universe Books, New York.

Meadows, D., Meadows, D. and Randers, J. 1992. *Beyond the Limits.* Chelsea Green Publishing Company, Post Mills, Vermont.

Munasinghe, M. and Cruz, W. 1994. *Economywide Policies and the Environment: Emerging Lessons from Experience.* The World Bank, Washington, DC.

Myers, N. 1993. *Gaia: An Atlas of Planet Management*, pp. 27–40. Anchor and Doubleday, Garden City, New York.

1994. "Tropical deforestation: rates and patterns." In K. Brown and D.W. Pearce, eds., *The Causes of Tropical Deforestation.* UCL Press, London.

1997. "Consumption: challenge to sustainable development or distraction?" *Science* 276:53–55.

Nordhaus, W. 1992. "Lethal Model II: *The Limits to Growth* Revisited." *Brookings Papers on Economic Activity* 2:1–43.

1994. *Managing the Global Commons: The Economics of Climate Change.* MIT Press, Cambridge, Massachusetts.

OECD. 1996. *Subsidies and the Environment: Exploring the Linkages.* Paris.

Office of Technology Assessment. 1988. *Using Desalination Technologies for Water Treatment.* U.S. Government Printing Office, Washington, DC.

Oldeman, R., Hakkeling, R. and Sombroeck, W. 1990. *World Map of the Status of Human-Induced Soil Degradation: An Explanatory Note.* United Nations Environment Program, Nairobi, Kenya.

Ostrom, E. 1990. *Governing the Commons*. Cambridge University Press, Cambridge, United Kingdom.

Page, J.M.Jr., Pearson, S.R. and Leland, H.E. 1976. "Capturing economic rent from Ghanaian timber." *Food Research Institute Studies* 15:25–51.

Panayotou, T. 1994. "The population, environment, and development nexus." In R. Cassen, ed., *Population and Development: Old Debates, New Conclusions*. Transaction Publishers, New Brunswick, New Jersey.

Panayotou, T. and Vincent, J. R. 1997. "Environmental regulation and competitiveness." In *The Global Competitiveness Report 1997*. World Economic Forum, Geneva.

Peters, P.E. 1994. *Dividing the Commons: Politics, Policy, and Culture in Botswana*. University of Virginia Press, Charlottesville, Virginia.

Pezzey, J. 1992. "Sustainable development concepts: an economic analysis." *World Bank Environment Paper No. 2*. The World Bank, Washington, DC.

Reed, D., ed. 1992. *Structural Adjustment and the Environment*. Westview Press, Boulder, Colorado.

Repetto, R. 1988. "Overview." In R. Repetto and M. Gillis, eds., *Public Policy and the Misuse of Forest Resources*, pp. 1–42. Cambridge University Press, Cambridge, United Kingdom.

    1995. *Jobs, Competitiveness, and Environmental Regulation: What are the Real Issues?* World Resources Institute, Washington, DC.

Repetto, R. and Gillis, M., eds. 1988. *Public Policies and the Misuse of Forest Resources*. Cambridge University Press, New York.

Repetto, R., Magrath, W., Wells, M., Beer, C. and Rossini, F. 1989. *Wasting Assets: Natural Resources in the National Income Accounts*. World Resources Institute, Washington, DC.

Repetto, R., Cruz, W., Solórzano, R. *et al.* 1991. *Accounts Overdue: Natural Resource Depreciation in Costa Rica*. World Resources Institute, Washington, DC.

Rosegrant, M., Agcaoili-Sombilla, M. and Perez, N. 1995. "Global food projections to 2020: implications for investment." *Food, Agriculture, and the Environment Discussion Paper No. 5*. International Food Policy Research Institute, Washington, DC.

Ruzicka, I. 1979. "Rent appropriation in Indonesian logging: East Kalimantan, 1972/73–1976/77." *Bulletin of Indonesian Economic Studies* 15(2):45–74.

Sachs, J. and Warner, A. 1995. "Natural resource abundance and economic growth." *Development Discussion Paper No. 517a*. Harvard Institute for International Development, Cambridge, Massachusetts.

Sagoff, M. 1997. "Do we consume too much?" *The Atlantic Monthly* (June).

Serageldin, I. 1995. *Toward Sustainable Management of Water Resources*. The World Bank, Washington, DC.

Shafik, N. and Bandyopadhyay, S. 1992. "Economic growth and environmental quality: time-series and cross-country evidence." *World Development Report Working Paper WPS 904*. The World Bank, Washington, DC.

Sizer, N. 1996. *Profit Without Plunder: Reaping Revenue from Guyana's Tropical Forests Without Destroying Them*. World Resources Institute, Washington, DC.

Sizer, N., and Rice, R. 1995. *Backs to the Wall in Suriname: Forest Policy in a Country in Crisis*. World Resources Institute, Washington, DC.

Solow, R. 1986. "On the intergenerational allocation of exhaustible resources."
    Scandinavian Journal of Economics 88(2):141–56.
Tobey, J. 1990. "The effects of domestic environmental policies on patterns of world
    trade: an empirical test." *Kyklos* 43(2):191–209.
Vincent, J. R. 1990. "Rent capture and the feasibility of tropical forest management."
    Land Economics 66(2):212–223.
    1992. "The tropical timber trade and sustainable development." *Science*
    256:1651–1655.
    1997. "Resource depletion and economic sustainability in Malaysia." *Environment
    and Development Economics* 2(1):19–37.
Vincent, J. R. and Fairman, D. 1995. "Multilateral consultations for promoting
    sustainable development through domestic policy changes." Paper presented at
    the Second Expert Group Meeting on Financial Issues of Agenda 21, held
    February 15–17, 1995, in Glen Cove, New York. Manuscript, Harvard Institute for
    International Development, Cambridge, Massachusetts.
Vincent, J. R. and Hartwick, J. 1997. "Accounting for the benefits of forest resources:
    concepts and experience." Report prepared for the FAO Forestry Department.
    FAO, Rome.
Vincent, J. R., Kaosa-ard, M., Worachai, L., Azumi, E. Y., Tangtham, N. and Rala, A. B.
    1995. "The economics of watershed management: a case study of Mae Taeng."
    Report prepared for the Natural Resources and Environment Program,
    Thailand Development Research Institute, Bangkok.
Vincent, J. R. and Panayotou, T. 1997. "Consumption: challenge to sustainable
    development . . . or distraction?" *Science* 276:53, 55–57.
Vincent, J. R., Panayotou, T. and Hartwick, J. 1997a. "Resource depletion and
    sustainability in small open economies." *Journal of Environmental Economics and
    Management* 33(3):274–286.
Vincent, J. R., Ali, R. M. and Associates. 1997b. *Environment and Development in a Resource-
    Rich Economy: Malaysia under the New Economic Policy.* Harvard University Press for
    Harvard Institute for International Development, Cambridge, Massachusetts.
Vincent, J. R. and Hadi, Y. 1993. "Malaysia." In National Research Council, *Sustainable
    Agriculture and the Environment in the Humid Tropics*, pp. 440–482. National
    Academy Press, Washington, DC.
Wallace, M. B. 1983. "Managing resources that are common property: from Kathmandu
    to Capitol Hill." *Journal of Policy Analysis and Management* 2(2):220–237.
Wattenberg, B. J. 1989. *The Birth Dearth.* Pharos Books, New York.
Weitzman, M. L. 1976. "Free access vs. private ownership as alternative systems for
    managing common property." *Journal of Economic Theory* 8:225–234.
Wheeler, D. and Mody, A. 1992. "International investment location decisions: the case of
    U.S. firms." *Journal of International Economics* 33:5–76.
World Bank. 1992. *1992 World Development Report: Environment and Development.*
    Washington, DC.
    1993. *Socio-Economic Time-Series Access and Retrieval System (STARS).* Version 3.0.
    Washington, DC.
    1997a. *Expanding the Measure of Wealth: Indicators of Environmentally Sustainable
    Development.* Washington, DC.
    1997b. *World Development Report 1997: The State in a Changing World.* Washington, DC.

NOTES

1. Appendix 1 in Pezzey (1992) gives eight pages of widely varying definitions.

2. Unless otherwise noted, in this chapter we define environmental degradation broadly, as including not only pollution but also the physical depletion of natural resources.

3. Economic surpluses, i.e., revenues minus costs.

4. Human and physical (human-made) capital account for the balance.

5. The social rate of time preference is given by the sum of the pure rate of time preference and the consumption rate of interest, which involves the elasticity of marginal utility (Arrow and Kurz 1970). The elasticity is higher when consumption is lower (i.e., when people are poorer).

6. This assumes that development policies do not suffer from an urban bias.

7. Based on analysis of production and consumption data in Table 4 of Annex 3 of FAO (1997).

8. Source given in the previous note.

9. For more sophisticated econometric analyses of income-pollution relationships for these and other parameters, see Grossman and Krueger (1991, 1995) and Shafik and Bandyopadhyay (1992).

10. For nitrogen, the coefficient was not significant at even the 5 percent level. Its t-statistic equaled only 1.17.

11. The negative correlation also might not hold for some other parameters. Indeed, in a regression for BOD (biochemical oxygen demand) in surface waters, the coefficient on the natural logarithm of per capita GDP was positive and statistically significant at the 5-percent level. Since higher levels of BOD represent greater pollution, this result is opposite the result for dissolved oxygen. The results for dissolved oxygen are statistically stronger, however, in that the sample for dissolved oxygen was larger in terms of both the number of observations (2682 vs. 1868) and the number of countries included (54 vs. 43), the fit of the regression equation was better ($R^2 = 0.172$ vs. 0.003), and the coefficient on per capita GDP was statistically more significant (t-statistic = 23.6 vs. 2.29). We thus place more confidence in the results for dissolved oxygen.

12. Rising educational levels are also important in this regard.

13. Two recent pairs of articles, by Myers (1997) and Vincent and Panayotou (1997) in *Science* and by Sagoff (1997) and Ehrlich *et al.* (1997) in *The Atlantic Monthly*, present contrasting views on the consumption issue.

14. Ben Wattenberg was one of the first demographers to draw attention to this issue, in his controversial book, *The Birth Dearth* (1989). Based on the views of critics as summarized in an appendix of the book, this issue was one of the less provocative ones he raised.

15. There is, however, a cluster of several dozen studies related to forest resources. Most focus exclusively on timber. For a review of these studies, see Vincent and Hartwick (1997).

16. Even if yields do not rise appreciably, world food supply models generally predict that (to quote one) "there will not be overwhelming pressure on aggregate world food supplies from rising populations and incomes" (Rosegrant *et al.* 1995, p. 30). The International Food Policy Research Institute's IMPACT model in fact predicts that real food prices will be lower in 2020 than in 1990 (Rosegrant *et al.* 1995).

17. Which incidentally require much less water per unit of energy generated than power plants fired by fossil fuels (Gleick 1993b).

18. The scope for technological advance is substantial. According to Gleick (1993b), the

currently "best" desalination plants use about 30 times the theoretical minimum amount of energy required.

19. A rational user would not select the stock associated with the *maximum* growth increment ("maximum sustained yield") unless harvest costs equal zero.

20. In fact, some economists have turned the Malthusian model on its head by attributing high rates of population growth in developing countries to poor parents' incentive to have many children to capture open access resources. See Dasgupta and Mäler (1991), Panayotou (1994), and Dasgupta (1995).

21. On community property rights, see Ostrom (1990).

22. Saying that the causes of environmental degradation are reversible is not the same as saying that environmental degradation itself is reversible. Improved policies help prevent future degradation, but past policies might leave a permanent mark on environmental quality.

23. For an analysis of this idea, see Vincent and Fairman (1995).

24. And, as the east Asian financial crises indicates, they are a mixed blessing if they are primarily short-term portfolio flows.

25. This value is based on the estimate reported in Table 5.1 of Nordhaus (1994), converted from 1989 to 1995 US dollars.

HERMAN E. DALY

3

Uneconomic growth: Empty-world versus full-world economics

That which seems to be wealth may in verity be only the gilded index of far-reaching ruin ... JOHN RUSKIN, *Unto This Last*, 1862.

Abstract

Uneconomic growth is growth that increases environmental and social costs by more than it increases production benefits. It is theoretically possible, yet appears anomalous within the neoclassical paradigm. Nevertheless, it is empirically likely that some northern countries have already entered a phase of uneconomic growth. Why does the dominant neoclassical paradigm make uneconomic growth seem anomalous, if not impossible? Why, by contrast, does uneconomic growth appear as an obvious possibility in the alternative paradigm of ecological economics? Although the neoclassical paradigm permits growth forever, it does not mandate it. Historically the mandate came because growth was the answer given to the major problems raised by Malthus, Marx, and Keynes. But when growth becomes uneconomic we must find new answers to the problems of overpopulation (Malthus); unjust distribution (Marx); and unemployment (Keynes). However, national policies required to deal with these three problems are undercut by "globalization" – the current ideological commitment to global economic integration via free trade and free capital mobility. The consequent erasure of national boundaries for economic purposes simultaneously erases the ability of nations independently to enforce policies for solving their own problems of overpopulation, unjust distribution, and unemployment. Many relatively tractable national problems are converted into one intractable global problem, in the name of "free trade", and in the interests of transnational capital.

Uneconomic growth in theory

Growth in GNP is so favored by economists that they call it "economic" growth, thus ruling out by terminological baptism the very possibility of "uneconomic" growth in GNP. But can growth in GNP in fact be uneconomic? Before answering this macroeconomic question let us consider the same question in the perspective of microeconomics – can growth in a microeconomic activity (firm production or household consumption) be uneconomic? Of course it can. Indeed, all of microeconomics is simply a variation on the theme of seeking the optimal scale or extent of each micro activity – the point at which increasing marginal cost equals declining marginal benefit, and beyond which further growth in the activity would be uneconomic because it would increase costs more than benefits.

But when we move to macroeconomics we no longer hear anything about optimal scale, nor about marginal costs and benefits. Instead of separate accounts of costs and benefits compared at the margin we have just one account, GNP, that conflates cost and benefits into the single category of "economic activity." The faith is that activity overwhelmingly reflects benefits. There is no macroeconomic analog of costs of activity to balance against and hold in check the growth of "activity", identified with benefits, and measured by GNP. Unique among economic magnitudes, GNP is supposed to grow forever.[1] But of course there really are costs incurred by GNP growth, even if not usually measured. There are costs of depletion, pollution, disruption of ecological life-support services, sacrifice of leisure time, disutility of some kinds of labor, destruction of community in the interests of capital mobility, takeover of habitat of other species, and running down a critical part of the inheritance of future generations. We not only fail to measure these costs, but frequently we even count them as benefits, as when we include the costs of cleaning up pollution as a part of GNP, and when we fail to deduct for depreciation of renewable natural capital and liquidation of nonrenewable natural capital.

There is no a priori reason why at the margin the costs of growth in GNP could not be greater than the benefits. In fact economic theory would lead us to expect that at some point. The law of diminishing marginal utility of income tells us that we satisfy our most pressing wants first, and that each additional unit of income is dedicated to the satisfaction of a less pressing want. So the marginal benefit of growth declines. Similarly, the law of increasing marginal costs tells us that we first make use of the most productive and accessible factors of production – the most fertile land, the most concentrated and available mineral deposits, the best workers – and only use the less productive factors as growth makes it necessary. Consequently, marginal costs of growth increase. When rising marginal costs equal falling marginal benefits then we are at the optimal level of GNP, and further growth would be uneconomic – would increase costs more than it increased benefits. Why is this simple extension of the basic logic of microeconomics treated as inconceivable in the domain of macroeconomics?[2] Mainly because microeconomics deals with the part, and expansion of a part is limited by the opportunity cost inflicted on the rest of the whole by the growth of the part under study. Macroeconomics deals with the whole, and the growth of the whole does not inflict an opportunity cost, because there is no "rest of the whole" to suffer the cost. Ecological economists have pointed out that the macroeconomy is not the relevant whole, but is itself a subsystem, a part of the ecosystem, the larger economy of nature.

Uneconomic growth in fact

One might accept the theoretical possibility of uneconomic growth, but argue that it is irrelevant for practical purposes since, it could be alleged, we are nowhere near the optimal scale. The benefits of growth might still be enormous and the costs still trivial at the margin. Economists all agree that GNP was not designed to be a measure of welfare, but only of activity. Nevertheless they assume that welfare is positively correlated with activity so that increasing GNP will increase welfare, even if not on a one-for-one basis. This is equivalent to believing that the marginal benefit of growth is greater than the marginal cost. This belief can be put to an empirical test. The results turn out not to support the belief.

Evidence for doubting the positive correlation between GNP and welfare in the United States is taken from two sources.

First Nordhaus and Tobin (1972) asked, "Is Growth Obsolete?" as a measure of welfare, hence as a proper guiding objective of policy. To answer their question they developed a direct index of welfare, called Measured Economic Welfare (MEW) and tested its correlation with GNP over the period 1929–1965. They found that, for the period as a whole, GNP and MEW were indeed positively correlated – for every six units of increase in GNP there was, on average, a four unit increase in MEW. Economists breathed a sigh of relief, forgot about MEW, and concentrated again on GNP. Although GNP was not designed as a measure of welfare, it was and still is thought to be sufficiently well correlated with welfare to serve as a practical guide for policy.

Some 20 years later John Cobb, Clifford Cobb, and I revisited the issue and began development of our Index of Sustainable Economic Welfare (ISEW) with a review of the Nordhaus and Tobin MEW. We discovered that if one takes only the latter half of their time series (i.e., the 18 years from 1947–1965) the positive correlation between GNP and MEW *falls* dramatically. In this most recent period – surely the more relevant for projections into the future – a six unit increase in GNP yielded on average only a one unit increase in MEW. This suggests that GNP growth at this stage of United States history may be a quite inefficient way of improving economic welfare – certainly less efficient than in the past.

The ISEW[3] was then developed to replace MEW, since the latter omitted any correction for environmental costs, did not correct for distributional changes, and included leisure, which both dominated the MEW and introduced many arbitrary valuation decisions. The ISEW, like the MEW, though less so, was positively correlated with GNP up to a point (around 1980) beyond which the correlation turned slightly negative. Neither the MEW nor ISEW considered the effect of individual country GNP growth on the *global* environment, and consequently on welfare at geographic levels other than the nation. Neither was there any deduction for legal harmful products, such as tobacco or alcohol, nor illegal harmful products such as drugs. No deduction was made for diminishing marginal utility of income resulting from growth over time. Such considerations would further weaken the correlation between GNP and welfare. Also, GNP, MEW, and ISEW all begin with Personal Consumption. Since all three measures have in common their largest single category, there is a significant autocorrelation bias, which makes the poor correlations between GNP and the two welfare measures all the more impressive.

Measures of welfare are difficult and subject to many arbitrary judgments, so sweeping conclusions should be resisted. However, it seems fair to say that for the United States since 1947, the empirical evidence that GNP growth has increased welfare is weak, and since 1980 probably nonexistent. Consequently, any impact on welfare via policies that increase GNP growth would also be weak or nonexistent. In other words, the "great benefit", for which we are urged to sacrifice the environment, community standards, and industrial peace, appears, on closer inspection, likely not even to exist.[4]

Uneconomic growth in two paradigms

Within the standard neoclassical paradigm uneconomic growth sounds like an oxymoron, or at least an anomalous category. You will not find it mentioned in any macroeconomics textbook. But within the paradigm of ecological economics it is an obvious possibility. Let us consider why in each case.

Neoclassical paradigm

The paradigm or preanalytic vision of standard neoclassical economics is that the economy is the total system, and that nature, to the extent that it is considered at all, is a sector of the economy – e.g., the extractive sector (mines, wells, forests, fisheries, agriculture). Nature is not seen as an envelope containing, provisioning, and sustaining the economy, but as one sector of the economy similar to other sectors. If the products or services of the extractive sector should become scarce, the economy will "grow around" that particular scarcity by substituting the products of other sectors. If the substitution is difficult, new technologies, in this view, will be invented to make it easy.

The unimportance of nature is evidenced, in this view, by the falling relative prices of extractive products generally and by the declining share of the extractive sector in total GNP. Beyond the initial provision of indestructible building blocks, nature is simply not important to the economy in the view of neoclassical economics.

That the above is a fair description of the neoclassical paradigm is attested by the elementary "principles of economics" textbooks, all of which present the shared preanalytic vision in their initial pages. This, of course, is the famous circular flow diagram, depicting the economy as consisting of a circular flow of value between firms and households – as an isolated system in which nothing enters from outside nor exits to the outside. There is no "outside", no environment.[5] Nature, if it is considered at all, is a sector of the macroeconomy. Further confirmation is found by searching the indexes of macroeconomics textbooks for any entries such as "environment", "nature", "depletion", or "pollution". The absence of such entries is nearly complete.

A personal experience confirmed to me even more forcefully just how deeply ingrained this preanalytic vision really is. I think it is worth taking the time to recount this experience, which had to do with the evolution of the World Bank's 1992 World Development Report (WDR), *Development and the Environment*.

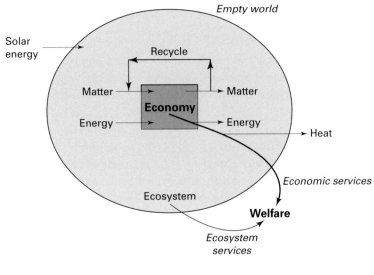

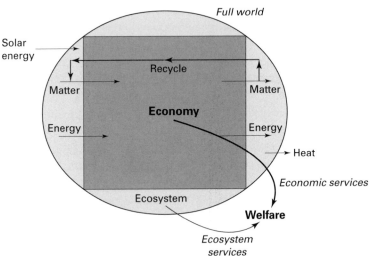

Figure 3.1 A "Macro" view of the macroeconomy.

An early draft of the 1992 WDR had a diagram entitled "The relation-ship between the economy and the environment". It consisted of a square labeled "economy", with an arrow coming in labeled "inputs" and an arrow going out labeled "outputs" – nothing more. I worked in the Envi-ronment Department of the World Bank at that time, and was asked to review and comment on the draft. I suggested that the picture was a good idea, but that it failed to show the environment, and that it would help to have a larger box containing the one depicted, and that the large box (or circle, perhaps) would represent the environment. Then the relation between the environment and the economy would be clear – specifically that the economy is a subsystem of the environment and depends on the environment both as a source of raw material inputs and as a sink for waste outputs. The text accompanying the diagram should explain that the environment physically contains and sustains the economy by regenerating the low-entropy inputs that it requires, and by absorbing the high-entropy wastes that it cannot avoid generating, as well as by supplying other systemic ecological services. Environmentally sustainable development could then be defined as development that does not destroy these natural support functions.

The second draft had the same diagram, but with an unlabeled box drawn around the economy, like a picture frame, with no change in the text. I commented that, while this was a step forward, the larger box really had to be labeled "environment" or else it was merely decorative, and that the text had to explain that the economy was related to the environment in the ways just described.

The third draft omitted the diagram altogether. There was no further effort to draw a picture of the relation of the economy and the environment. Why was it so hard to draw such a simple picture?

By coincidence a few months later the Chief Economist of the World Bank, under whom the 1992 WDR was being written, happened to be on a review panel at the Smithsonian Institution discussing the book *Beyond the Limits* (Meadows *et al.*). In that book was a diagram showing the relation of the economy to the ecosystem as subsystem to total system, identical to what I had suggested. In the question-and-answer time I asked the Chief Economist if, looking at that diagram, he felt that the issue of the physical size of the economic subsystem relative to the total ecosystem was important, and if he thought economists should be asking the question, "What is the optimal scale of the macro economy relative to the environment that supports it?" His reply was short and definite, "That's not the right way to look at it", he said.

Reflecting on these two experiences has strengthened my belief that the difference truly lies in our "preanalytic vision" – the way we look at it. My preanalytic vision of the economy as subsystem leads immediately to the questions: How big *is* the subsystem relative to the total system? How big *can it be* without disrupting the functioning of the total system? How big *should it be*, what is its optimal scale, beyond which further growth in scale would be uneconomic – would increase environmental costs more than it increased production benefits? The Chief Economist had no intention of being sucked into these subversive questions – that is not the right way to look at it, and any questions arising from that way of looking at it are simply not the right questions.

That attitude sounds rather unreasonable and peremptory, but in a way that had also been my response to the diagram in the first draft of *Development and the Environment* showing the economy receiving raw material inputs from nowhere and exporting waste outputs to nowhere. "That is not the right way to look at it", I said, and any questions arising from that picture, say, how to make the economy grow as fast as possible by speeding up throughput from an infinite source to an infinite sink, were not the right questions. Unless one has in mind the preanalytic vision of the economy as subsystem, the whole idea of sustainable development – of an economic subsystem being sustained by a larger ecosystem whose carrying capacity it must respect – makes no sense whatsoever. It was not surprising therefore that the 1992 WDR was incoherent on the subject of sustainable development, placing it in solitary confinement in a half-page box where it was implicitly defined as nothing other than "good development policy". It is the preanalytic vision of the economy as a box floating in infinite space that allows people to speak of "sustainable *growth*" (quantitative expansion) as opposed to "sustainable *development*" (qualitative improvement). The former term is self-contradictory to those who see the economy as a subsystem of a finite and nongrowing ecosystem. The difference could not be more fundamental, more elementary, or more irreconcilable.

Ecological economics paradigm

This story of course leads to a consideration of the alternative paradigm, that of ecological economics within which uneconomic growth is an obvious concept. The big difference is to see the economy as a subsystem of the natural ecosystem.

The neoclassical "evidence" for the unimportance of nature (falling relative price of many natural resources, and small share of the extractive sector in GNP) is seen quite differently in the ecological economics paradigm. In an era of rapid extraction of resources their short-run supply will of course be high and their market price consequently will be low. Low resources prices are not evidence of nonscarcity and unimportance,

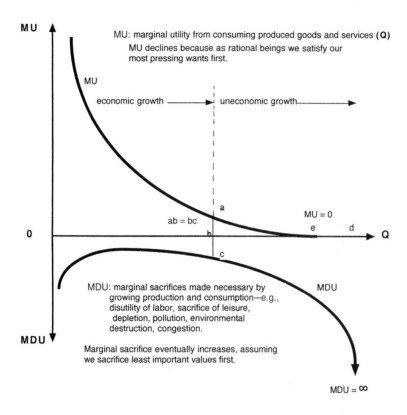

MU

MU: marginal utility from consuming produced goods and services **(Q)**
MU declines because as rational beings we satisfy our
most pressing wants first.

MU

economic growth ⟶ uneconomic growth ⟶

a
ab = bc   MU = 0
h   e   d
0   Q

c

MDU: marginal sacrifices made necessary by   MDU
growing production and consumption—e.g.,
disutility of labor, sacrifice of leisure,
depletion, pollution, environmental
destruction, congestion.

MDU

Marginal sacrifice eventually increases, assuming
we sacrifice least important values first.

MDU = ∞

Limits to Growth of Macroeconomy

b = economic limit;   MU = MDU; (maximum net positive utility)

e = futility limit; MU = 0;   (consumer satiation)

d = catastrophe limit; MDU = ∞; (ecological disaster)

Figure 3.2 Jevonian view of limits to growth of macroeconomy.

but rather evidence of rapid use and increasing technological dependence
on a large throughput of cheap resources. As for the neoclassical claim
that the small percentage of GNP arising from the extractive sector indicates its unimportance, one might as well claim that a building's foundation is unimportant because it represents only five percent of the height
of the skyscraper erected above it. GNP is the sum of value *added* by labor
and capital. Resources are *that to which value is added* – the base or foundation upon which the skyscraper of added value is resting. A foundation's

importance does not diminish with the growth of the structure that it supports! If GNP growth resulted only from increments in value added to a nongrowing resource throughput, then it would likely remain *economic* growth. Such a process of qualitative improvement without quantitative increase beyond environmental capacity is what I have elsewhere (Daly 1966) termed development without growth, or "sustainable development". But that is not what happens in today's world.

What happens, according to ecological economics, is that the economy grows by transforming its environment (natural capital) into itself (manmade capital). The optimal extent of this physical transformation (optimal scale of the economy) occurs when the marginal cost of natural capital reduction is equal to the marginal benefit of manmade capital increase. This process of transformation takes place within a total environment that is finite, nongrowing, and materially closed. There is a throughput of solar energy that powers biogeochemical cycles, but that energy throughput is also finite and nongrowing. As the economic subsystem grows it becomes larger relative to the total system, and therefore must conform itself more to the limits of the total system – finitude, nongrowth, and entropy. Its growth is ultimately limited by the size of the total system of which it is a part, even under neoclassical assumptions of easy substitution of manmade for natural capital. But if manmade and natural capital are complements rather than substitutes, as ecological economics claims, then expansion of the economic subsystem would be much more stringently limited. There would be no point in expanding manmade capital beyond the capacity of remaining natural capital to complement it. What good are more fishing boats when the fish population has disappeared? The fish catch used to be limited by number of fishing boats (manmade capital) but is now limited by the remaining populations of fish in the sea (natural capital).

When factors are complements the one in short supply is *limiting*. If factors are substitutes then there cannot be a limiting factor. Economic logic says that we should focus attention on the limiting factor by (a) in the short run maximizing its productivity; and (b) in the long run investing in its increase. This is a *major* implication for economic policy – economize on and invest in natural capital. Economic logic stays the same, but the identity of the limiting factor has gradually changed from manmade to natural capital – e.g., from fishing boats to remaining fish in the sea; from saw mills to remaining forests; from irrigation systems to aquifers or rivers; from oil well drilling rigs to pools of petroleum in the ground;

from engines that burn fossil fuel to the atmosphere's capacity to absorb CO_2, etc.

Viewed from the perspective of ecological economics even the usual neoclassical assumption of easy substitution of manmade for natural capital provides no argument for growth – at least not when growth is viewed as the transformation of natural into manmade capital. If manmade capital substitutes for natural capital, then natural capital substitutes for manmade capital. Substitution is reversible. If our original endowment of natural capital was a good substitute for manmade capital, then why, historically, did we go to the trouble of transforming so much natural capital into manmade capital? Neoclassical believers in easy substitution have no good answer. The problem does not arise for ecological economists because they affirm from the beginning that natural and manmade capital are basically complements and only marginally substitutes.

The smaller the optimal scale of the economy, the greater are (a) the degree of complementarity between natural and manmade capital, (b) our desire for direct experience of nature, and (c) our estimate of both the intrinsic and instrumental value of other species. The smaller the optimal scale of the economy, the sooner its physical growth becomes uneconomic.

From permitting growth, to mandating growth, to limiting growth

The neoclassical paradigm permits growth forever but does not mandate it. Historically, what pushed the growth-forever ideology was the answer given to the problems raised by Malthus, Marx, and Keynes. Growth was the common answer to all three problems. Overpopulation, unjust distribution, and involuntary unemployment would all be solved by growth. Overpopulation would be cured by the demographic transition initiated by growth. Unjust distribution of wealth between classes would be rendered tolerable by growth, the rising tide that lifts all boats. Unemployment would yield to increasing aggregate demand that merely required that investment be stimulated, which of course implies growth. Continuing this time-honored tradition the World Bank's 1992 WDR argued that more growth was also the solution to the environmental problem. But of course the assumption in all cases was that growth was economic, that it was making us richer rather than poorer. But now

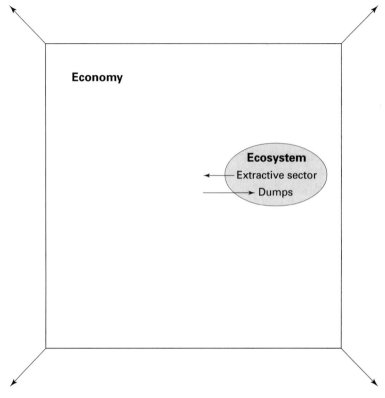

Figure 3.3 Ecosystem as subsystem of macroeconomy.

growth is becoming uneconomic. Uneconomic growth will not sustain the demographic transition and cure overpopulation. Neither will it help redress unjust distribution, nor cure unemployment. Nor will it provide extra wealth to be devoted to environmental repair and clean-up.

We now need more radical and direct solutions to the problems of Malthus, Marx, and Keynes: population control to deal with overpopulation, redistribution to deal with excessive inequality; and ecological tax reform to raise resource productivity and employment. These must be national policies. It is utopian (or dystopian) to think of them being carried out by a world authority. Many nations have made progress in controlling their population growth, in limiting domestic income inequality, in reducing unemployment. They have also improved resource productivity by internalizing environmental and social costs into prices. Nations must enact policies to stop uneconomic growth. But their efforts in this regard are undercut by the ideology of globalization.

Globalization as stimulus to uneconomic growth

Global economic integration by free trade and free capital mobility effectively erases the policy significance of national boundaries, turning the federated community of nations into a cosmopolitan noncommunity of globalized individuals. Some of these "individuals" are giant transnational corporations, but legally treated as fictitious individuals. Nations can no longer internalize environmental and social costs in the interests of resource efficiency, because capital is free to produce elsewhere and still sell its product in the market whose social controls it just escaped. In like manner capital escapes higher wages and taxes of any kind, in particular taxes aimed at redistributive policies that redress excessive inequality and poverty.

Just as it is hard to imagine a country internalizing its external costs when forced to trade freely with countries that do not, so too it is hard to imagine any country continuing to limit its birth rate when the results of overpopulation in other countries spill over into it. Whether capital moves to overpopulated low-wage countries, or poor workers move to the high-wage country, the result is the same – a competitive bidding down of wages to the detriment of countries that have followed a high-wage policy by limiting their numbers and more equally distributing their wealth. The laboring class in the low-wage country gains in terms of number employed, though not usually in terms of increased wages because of overpopulation and continuing high demographic growth, especially among the working class. The capitalist class in the high-wage country gains from lower wage costs at home. The big losers are workers in the (formerly) high-wage countries. Indeed, with low wages now a competitive advantage in attracting capital, we might expect policies aimed at increasing the supply of labor in previously high-wage countries. Already the *Wall Street Journal* calls for easy immigration into the U.S.A. Before long someone will likely advocate higher birth rates for the laboring class in high-wage countries as a solution for alleged "labor shortages". Furthermore, with falling real wages and disappearing social security, it is possible that we might even have a reversion to larger working class families in search of security and community – a reverse demographic transition.

Under globalization, each country seeks to expand beyond the limits of its own ecosystem and market by growing into the ecological and economic space of all other countries, as well as into the remaining global commons. Globalization operates by standards-lowering competition to bid down wages, to externalize environmental costs, and reduce social overhead charges for welfare, education, and other public goods. It is far worse than an unrealistic global dream – it actively undercuts the ability of nations to continue to deal with their own problems of overpopulation, unjust distribution, unemployment, and external costs. It converts many relatively tractable national problems into a single intractable global problem.

Globalization via export-led growth is the new philosopher's stone of the IMF–IBRD–WTO alchemists.[6] Nations can all turn their lead into gold by free trade. With the revival of alchemy comes a return to the logic of Mercantilism: wealth is gold, and the way for countries without mines to get gold is to export more goods than they import, and receive payment for the difference in gold – the alchemy of trade. The way to export more than you import is to reduce wages and to externalize social and environmental costs, because that keeps prices of your exports competitive. Low wages also prevent your laboring class majority from importing. The way to keep wages low is to have an oversupply of labor. An oversupply of labor can be attained by easy immigration and high birth rates among the working class. Globalization requires, therefore, that for a nation to be rich, the working-class majority of its citizens must be poor, increase in number, and live in a deteriorating environment. Behind these absurdities is the deeper contradiction that under globalization it no longer makes sense to speak of "nations" (only corporations), nor of "citizens" (only employees).

Truly, globalization is accelerating the shift to uneconomic growth, a time when, as John Ruskin foresaw, "That which seems to be wealth may in verity be only the gilded index of far-reaching ruin . . ."

REFERENCES

Cobb, C., Halstead, T, and Rowe, T. October 1995. "If the GDP is up, why is America down?" *Atlantic Monthly*: 59–78.

Cobb, C.W., Cobb, J.B. Jr. and the Human Economy Center 1994. T*he Green National Product*. University Press of America, New York.

Daly, H. and Cobb, J. 1989. *For the Common Good*. Beacon Press, Boston. 2nd edition, 1994.

Daly, H. 1966. *Beyond Growth: The Economics of Sustainable Development*. Beacon Press, Boston, Massachusetts.

Max-Neef, M. 1995. "Economic growth and Quality of Life: A Threshold Hypothesis."
    Ecological Economics 15: 115–118.

Meadows, D., Meadows, D. and Randers, J. 1992. *Beyond the Limits.* Chelsea Green
    Publishing Company, Post Mills, Vermont.

Nordhaus, W. and Tobin, J. 1972. "Is Growth Obsolete?" In National Bureau of
    Economic Research, *Economic Growth.* Columbia University Press, New York.

NOTES

1. Although macroeconomists see no limits on the *size* of GNP, they have recognized a
limit on its *rate of growth* in the form of inflation that results as the economy approaches
full employment. This is seen more as an institutional limit, than a biophysical one.

2. Of course if wants and technology change, as surely they do, then the optimal level of
GNP will change. But there would then be another optimum beyond which growth
would again be uneconomic. It is gratuitous to assume that changes in wants and
technology will always be of a kind that results in a larger optimal GNP. The growth-
forever paradigm has been saved in practice by: increasing focus on insatiable relative
wants to the neglect of satiable absolute wants, aggressive advertising, increasing debt,
and falling monetary costs of production attained by externalization of the real costs of
more powerful and dangerous technologies.

3. For critical discussion and the latest revision of the ISEW, see Cobb *et al.* (1994). For a
presentation of the ISEW see Appendix of Daly and Cobb (1989). See also Cobb *et al.*
(1995).

4. For further evidence from other countries, see Max-Neef (1995).

5. The economic animal has neither mouth nor anus – only a closed-loop circular gut –
the biological version of a perpetual motion machine!

6. To these acronyms we may soon have to add MAI (Multilateral Agreement on
Investment), a proposal currently being pushed in the OECD as a first step toward
world agreement. This agreement would impose *de jure* what is now being achieved *de
facto* by standards-lowering competition to attract mobile capital – namely the erasure
of any distinction between national and foreign investment.

4
—————

Population and consumption: From more to enough

The world is in the midst of a transition to a more crowded and more consuming, warmer and more stressed, more connected, yet diverse and divided world. Population growth, greater consumption, environmental and technological change, and increasing connectedness and diversity are powerful trends that will continue through the early decades of the next century. A hopeful vision for such a transition is one in which the many more people of the next half century meet their wants and needs in sustainable ways that move away from ones that degrade the planet's life support systems towards ones that sustain or restore them. Central to such a sustainability transition will be, for both population and consumption, a transition from more to enough. But in this comparative review, I find a profound asymmetry between what is known and understood about population and what is known and understood about consumption. Learning to manage consumption at a level equivalent to population may be the major challenge of a sustainability transition (Kates 1994, 1995, Myers 1997a, 1997b, Vincent and Panayotou 1997).

Population vs. consumption

A recent report from the National Research Council begins as follows:

> For over two decades, the same frustrating exchange has been repeated countless times in international policy circles. A government official or scientist from a wealthy country would make the following argument:
>
> > The world is threatened with environmental disaster because of the depletion of natural resources (or climate change, or the loss of biodiversity), and it cannot continue for long to support its

rapidly growing population. To preserve the environment for future generations, we need to move quickly to control global population growth, and we must concentrate the effort on the world's poorer countries, where the vast majority of population growth is occurring.

Government officials and scientists from low-income countries would typically respond this way:

If the world is facing environmental disaster, it is not the fault of the poor, who use few resources. The fault must lie with the world's wealthy countries, where people consume the great bulk of the world's natural resources and energy and cause the great bulk of its environmental degradation. We need to curtail overconsumption in the rich countries which use far more than their fair share, both to preserve the environment and to allow the poorest people on earth to achieve an acceptable standard of living

(STERN *et al.*, 1997, p. 1)

This view from the summit is correct, but it is not necessarily the view from below. Survey research in both industrialized and developing countries show that there is equal, and perhaps even greater concern, in developing countries for their environmental problems (Bloom 1995)

Nor is it necessarily the view of scientists. For 25 years now there has been a general consensus among environmental scientists that growth in population, in wealth, and in technology are jointly responsible for environmental problems (Dietz and Rosa 1994). This has become enshrined in a useful, albeit overly simplified, identity known as IPAT (I = Population×Affluence×Technology). In this identity, various forms of environmental or resource impacts are a function of population, wealth (usually income per capita), and the impacts per unit of income as determined by the technology and institutions in use. At various times and places, change in population, affluence, or technology are of greater or lesser importance, providing both evidence and anecdote to support the differing views at the summit.

The differing views at the summit and the frustration of developing country politicians and scientists alike are fueled by the profound asymmetry in how much we know about population and how little we know about consumption. Each differs in how it is defined, in the available data and known trends, in the theories that provide understanding, and in the policies that can be supported by such understanding.

Population

The definition of population and the elements of population growth are well understood, if not the forces driving them. The data are comparatively excellent and provide evidence of a demographic transition, as both description and theory, as well as support for policies designed to accelerate a global transition.

Definitions

Population begins with people and their key events of birth, death, and location. At the margins, there is some debate: when does life begin and end or what to do about temporary residence, but little debate between. Thus change in the population of the world, is the simple arithmetic of adding births, subtracting deaths, and for particular places, adding immigrants, and subtracting emigrants. While whole subfields of demography are devoted to the arcane details of these additions and subtractions, population for almost all places is known within 20 percent and for countries with modern statistical services within 3 percent – better estimates than for any other living thing and for most other environmental concerns.

Trends

The excellence of current population data has been extended to the past with great ingenuity. Deevey was the first to argue for three great population surges (Deevey 1960) and subsequent studies have supported his estimates (Figure 4.1). The tool-making or cultural revolution, which lasted about a million years, eventually supported upwards of 5 million people. The original "green" or agricultural revolution that followed the domestication of plants and animals and the invention of agriculture fueled population growth a hundred-fold over the next 8000 years. Now we are in the midst of a third great surge begun in the industrial revolution.

Current world population is about 5.8 billion people, growing at a rate of 1.5 percent per annum with about 80 million people added each year. About 80 percent or 4.6 billion people live in the less developed areas of the world with 1.2 billion living in industrialized countries (Population Reference Bureau 1998). The peak growth rate of all history, about 2.1 percent, occurred in the early 1960s and the peak population increase of 87 million occurred in the late 1980s (Kates 1995). Population

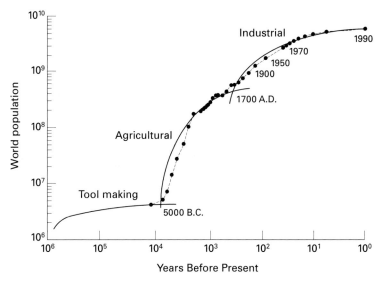

Figure 4.1 World population with three growth pulses (Kates, 1997, Figure 3, p. 38).

is now projected by the UN to be 9.4 billion in 2050, according to the medium-fertility variant (the one usually considered to be most likely) or as high as 11.1 billion or as low as 7.7 billion and to virtually level by 2100 completing the third great surge in population numbers (United Nations 1998).

Theories

There is a general description of how birth rates and death rates change over time, a process called the demographic transition (Davis 1990). In such a transition, deaths decline more rapidly than births and in that gap population grows rapidly but eventually stabilizes as the birth decline matches or even exceeds the death decline. While the general description of the transition is widely accepted, there is much that is debated as to cause and detail.

It was first studied in the context of Europe, where in the space of two centuries, societies went from a condition of high births and high deaths to the current situation of very low births and low deaths. While the decline in births lagged behind the decline in deaths, both changes took place during the period of industrialization as European society moved from an agrarian base to an urban-industrial base. Thus most scholars associate the decline in births and deaths with such periods of modernization or development (Davis 1990).

The rapid decline in deaths is not truly understood (Walter and Schofield 1989) and is variously attributed to some combination of improved nutrition, living standards, disease prevention, and medical treatment. While it is easy to presume that the decline was the result of improved medical care and prevention in the form of immunization, the death rate began to decline long before the science and practice of modern medicine evolved (McKewon 1976).

Scholars also differ about which elements of development would encourage the decline in births – whether it was the changing economics and usefulness of family labor, the improved security of family size with reduced infant and child deaths, or greater knowledge and interest in birth control resulting from education, particularly of women. Fertility began its decline in most European countries between 1880 and 1930 (Knodel and van de Walle 1976), but France began its decline before the industrial revolution even began. Births declined in countries where the development process was well underway (Belgium, England, Germany, Switzerland) as well as in countries predominantly agrarian and with large numbers of illiterates (Bulgaria, Hungary, Italy, and Spain). And in all European countries infant mortality remained high when the decline began. Thus while the empirical facts of the demographic transition are clear, the explanations of causation are not, even in this, the best studied historic case.

We also know that we are now in the midst of a global transition that unlike the European transition is much more rapid. Both births and deaths have dropped much faster than experts expected and history foreshadowed. It took a hundred years for deaths to drop in Europe compared to the drop in 30 years in the Third World. Today the global transition to required stability is more than halfway there, between the average of 5 children born to each woman at the post World War II peak of population growth and the 2.1 births required to achieve eventual zero population growth. Current births average 2.9 (Population Reference Bureau 1998). The death transition is more advanced, life expectancy having grown about three-quarters of the transition between a life expectancy at birth of 40 years to one of 75, and is currently at 66 years (Population Reference Bureau 1998). Indeed the current rates of decline in births outpace the estimates of the demographers, the UN having reduced its latest medium expectation of global population in 2050 to 9.4 billion, a reduction of almost 5 percent from the 1994 projection! (United Nations 1998).

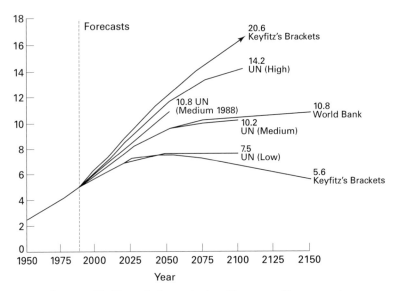

Figure 4.2 World population projections (Kates, 1997, Figure 7, p. 51).

However, the range of population projections has not changed for over two decades (Figure 4.2). The United Nations (1992) and the World Bank (Bos *et al.* 1994) have projected a world population of between 7–15 billion with a likely number of 10–11 billion that stabilizes within the next two centuries (Figure 4.2). But this broad range may be too small (McNicoll 1992, Cohen 1995), and Keyfitz and others argue that there is a good chance that in the year 2100, global population will actually fall somewhere between 5 and 20 billion people (Keyfitz 1981, Stoto, 1983, Lee 1989). Yet despite such differences, most demographers seem to believe that the momentum of the demographic transition is well underway, and many believe that there is sufficient empirical and theoretical understanding to design policies to accelerate it.

Policies
In one such effort, Bongaarts (1994) has decomposed the projected growth of the next century by three major causes (Figure 4.3) and has sought to accelerate the demographic transition with policies that would encourage more rapid fertility declines. The population of the developing world (using earlier projections) was expected to reach 10.2 billion by 2100. Of this growth, 1.9 billion will be born to the 120 million married women (and the many unmarried women) who either want fewer children or to

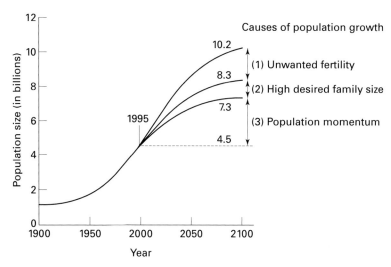

Figure 4.3 Decomposition of population growth in developing countries by major cause (Bongaarts 1994, Figure 4).

space them better. Thus, in theory, meeting all of the unmet need for contraception could reduce this total to 8.3 billion. A basic strategy for doing so links voluntary family planning with other reproductive and child health services (Bongaarts and Bruce 1997).

A further billion children reflect the current desires for large families in many parts of the world. For this, Bongaarts would seek to change the costs and benefits of child rearing so that more parents will recognize the value of smaller families, while simultaneously increasing the investment in children. An immediate and effective implementation of measures to bring down desired family size to replacement fertility would reduce the population size in 2100 further to 7.3 billion. A basic strategy for doing so accelerates three trends that have been shown to lead to lower desired family size: the survival of children, their education, and improvement in the economic, social, and legal status of girls and women (Bongaarts and Bruce 1997).

The remaining growth – from 4.5 billion today to 7.3 billion in 2100 – would be due to the momentum of population growth arising from the large number of prospective parents that are alive today and the short interval between generations. Bongaarts would slow this momentum by increasing the age of child bearing, primarily by improving the secondary education and income-generating opportunities for adolescent girls, and

---

Box 4.1 What's Consumption?
Physicist: "What happens when you transform matter/energy"
Ecologist: "What big fish do to little fish"
Economist: "What consumers do with their money"
Sociologist: "What you do to keep up with the Joneses"

---

by addressing neglected issues of their sexuality and reproductive behavior (Bongaarts and Bruce 1997).

For each of the specific policy directions, there are extensive examples of current implementation in developing countries of varied wealth, ethnic makeup, and political persuasion (Bongaarts and Bruce 1997). Using such "best practice" examples and applying them just to those countries where they may be most rapidly and effectively implemented could, in my judgment, lead to a reduction in projected population – up to 10 percent, or a billion, fewer people.

Consumption

In contrast to population, where people, their births, and deaths, are relatively well-defined biological events, there is no consensus as to what consumption consists of. With the exception of energy, there are limited data, suggestive trends but little theory, and at best policy directions.

Definitions

Stern (1997) has described the different ways physics, economics, ecology and sociology view consumption (Box 4.1). For physicists, matter and energy cannot be consumed, so consumption is conceived as transformations of matter and energy with increased entropy. For economists, consumption is what consumers do with their money, spending on consumer goods and services as distinguished from their production and distribution. For ecologists, consumption is what big fish do to little fish, obtaining energy and nutrients by eating something else, mostly green plants or other consumers of green plants. And for some sociologists, consumption is a status symbol – keeping up with the Joneses – when individuals and households use their incomes to increase their social status through certain kinds of purchases.

The Royal Society of London and the U. S. National Academy of

Sciences (1997) issued a joint statement on consumption, having previously done so on population. They chose a variant of the physicist's definition, saying:

> Consumption is the human transformation of materials and energy. Consumption is of concern to the extent that it makes the transformed materials or energy less available for future use, or negatively impacts biophysical systems in such a way as to threaten human health, welfare, or other things people value.
>
> (p. 684)

This Society/Academy view has a two-handed virtue. On the one hand it is more holistic and fundamental than the other definitions; on the other hand it is more focused, turning our attention to that which is environmentally damaging. In what follows, I use it as a working definition with one modification. I would add information to energy and matter, thus completing the triad of the biophysical and ecological basics that support life. And information may also be important in substituting for environmentally damaging consumption.

Trends

For consumption as the transformation of energy, materials, and information there are limited data as to trends. There is relatively good global knowledge of energy transformations due in part to the common units of conversion between different technologies (Nakicenovic 1996). Between 1950 and 1992, global energy production and use increased almost fourfold (Worldwatch Institute 1998).

For material transformations there are no aggregate data in common units on a global basis, only for some specific classes of materials including materials for energy production, construction, industrial minerals and metals, agricultural crops, and water (World Resources Institute *et al.* 1996). Calculations of material use by volume, mass, or value lead to different trends.

For the U.S.A. there are good trend data for major classes of materials. For example, trends in annual per capita use of physical structure materials (construction and industrial minerals, forestry products), show a logistic growth pattern: modest doubling between 1900 and the depression of the 1930s (from 2 to 4 metric tons), followed by a steep quintupling with economic recovery until the early 1970s (from 2 to 11 tons), followed by a leveling off since then with fluctuations related to economic downturns (Wernick 1996).

There is also now an aggregate analysis of current material production and consumption in the U.S.A. by mass, which averages well over 60 kilos per person per day (excluding water). Three-quarters of the material flow is split between energy and related feedstock conversion (38 percent) and minerals for construction (37 percent), with the remainder as industrial minerals (5 percent), metals (2 percent), and products of fields (12 percent) and forest (5 percent) (Wernick and Ausubel 1995).

For information, a massive effort is underway to catalog biological (genetic) information (B. Alberts personal communication). Currently, we have the complete genomes of perhaps 100 viruses, 10 bacteria, baker's yeast and a worm. Major efforts are underway to sequence the genomes of various disease microbes, a fly, a plant, mice, and humans. In contrast to the molecular detail, the number and diversity of organisms is unknown with a conservative estimate of species being $\sim 10^7$, of which only an estimated 10^6 have been described, and only $\sim 10^5$ are well known (Pimm *et al.* 1995).

Neither concepts nor data are available on most cultural information, although there are some gross measures. For example, the number of languages in the world continues to decline while the number of messages expands exponentially. Only 6000 languages remain of the estimated 10–15 000 that may have existed simultaneously at any one time and half of these are no longer learned by children and may go extinct in the coming century (Kane 1997). By contrast the number of communications in the form of *international* telephone traffic, calls just between countries, more than tripled in a single decade (1985–95), is currently averaging ten minutes a year for each person on earth, and is growing at a rate of 12 percent per annum. By the end of the century it is expected that there will be 1.2 billion telephone subscribers, including 300 million mobile links (Telegeography 1997).

Trends and projections in agriculture, energy, and economy can serve as surrogates for more detailed data on energy and material transformation. For comparative purposes they are expressed as multiples of experienced or projected population growth in Table 4.1. From 1950 to the early 1990s, world population more than doubled ($\times 2.2$), food as measured by grain production almost tripled ($\times 2.7$) energy more than quadrupled ($\times 4.4$) and economy quintupled ($\times 5.1$) (Brown *et al.* 1994). This 43-year record can be compared to two sets of projections or scenarios that reflect the continuation of current trends or as some note "business as usual". They differ somewhat by time period and most importantly by their assumptions as to meeting future needs.

Table 4.1. *Actual and projected change in world population, food, energy, and economy*

	Actual 1950–1993	Conventional Development 1990–2050	Minimal Necessities 1975–2075
Population	×2.2	×1.9	×2.5
Food (Grain)	×2.7	×2.0	×3.75
Energy	×4.4	×2.7	×5.9
Economy (GDP)	×5.1	×4.4	×8.0

Sources: For actual – Brown *et al*. 1994; for conventional development – Raskin *et al*. 1995; and for minimal necessities – Anderberg, 1989.

The "conventional development paradigm" assumes the continuity of "the socio-economic arrangements, values, and lifestyles that evolved during the industrial era" with no greater effort in the future to reduce global inequities than is current today. In this 60 year projection, an almost doubling of population (×1.9), finds a doubling of agriculture (×2.0), almost a tripling of energy (×2.7), and a quadrupling of the economy (×4.4) (Raskin *et al*. 1995). An older study focused on the century 1975–2075, choosing optimistic projections well within the range of the "conventional wisdom" but intended to provide an enlarged consumption that allows for meeting minimal necessities and lessening global inequities without radical restructuring of the world's economies. A century long more-than-doubling of population (×2.5) required roughly a quadrupling of agriculture (×3.75), a sextupling of energy (×5.9), and an octupling of the economy (×8.0), if varied and nutritious diets, industrial products, and regular jobs were to be in reach of most of the ten billion people (Anderberg 1989).

Theories

Thus whether future growth in population and consumption, grows only arithmetically (2, 3, 4, 5 times) or geometrically (2, 4, 6, 8 times), both history and future scenarios predict growth rates of consumption well beyond population. But there is an attractive similarity between a demographic transition that moves over time from high births and high deaths to low births and low deaths with an energy, materials, and information transition. In this transition, societies will use increasing amounts of

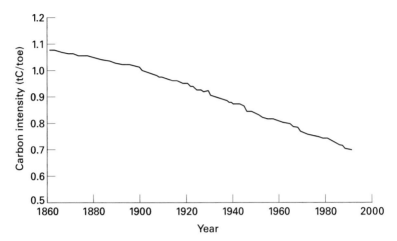

Figure 4.4 Carbon intensity of primary energy (Nakicenovic 1997, Figure 2, p. 77).

energy and materials as consumption increases, but over time the energy and materials input per unit of consumption decreases and information substitutes for more material and energy inputs.

Wernick (1996) describes a hypothesis in which over time materials (including those used for energy production) that require less mass to deliver a unit of given service are substituted for heavier materials, with the aggregate reductions in the amount of materials needed to serve economic functions resulting in dematerialization. The hypothesis argues that as developing nations industrialize, their materials use peaks in shorter times than it did historically in industrial countries, but their overall consumption saturates at lower levels as they can take advantage of the trial and error experience of the industrialized countries in material use.

There are some encouraging signs for such a transition in phenomena variously labeled as decarbonization and dematerialization, where over time, more than a century in the case of carbon (Figure 4.4) there exists a trend of making more with less (Nakicenovic 1997, Wernick et al. 1997). But the evidence for dematerialization currently exists only for industrialized countries and for specific materials.

The logistic trend in the per capita use of physical structural materials in this century (described above) suggests a kind of materials transition with its use now leveling off. Trends in specific material use per value of gross product in the United States (Figure 4.5), and for that matter in

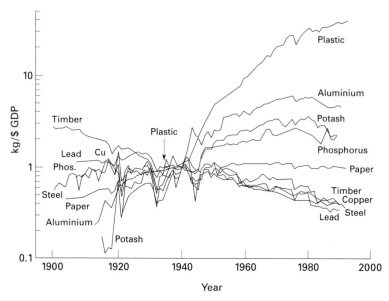

Figure 4.5 U. S. materials intensity of use (Wernick *et al.* 1997, Figure 1, p. 140).

industrial countries generally, show we are clearly using less wood, steel, lead and copper, and cement per unit of economic production, but more plastic, aluminum, and fertilizers (although these have peaked and are beginning to lessen). Surprisingly, despite the computer and television revolutions, the use of paper remains constant.

The flows of energy and materials, essential to maintaining economic well-being and quality of life, increase in the aggregate and decrease per unit of product and service. Improvements in technology and substituting information for energy and materials will continue to drive increased energy efficiency (including decarbonization) and dematerialization per unit of product or service. At the same time, the demand for products and services continues to increase and the overall consumption of energy and most materials more than offsets the efficiency and productivity gains.

Policies
The type of understanding that permits the decomposition of future population growth in ways that would permit us to slow it down is lacking in terms of consumption. Consumption as the aggregate of energy and material and information transformations is too complex for simple decomposition even if we had all the data and trends. Yet, it may be

possible to decompose major components of such transformations and a recent workshop explored existing studies for four consumption needs of great significance for both human life and the environment: food (Bender 1997), energy (Schipper *et al.* 1997), forest products (Wernick *et al.* 1998), and water (Gleick 1993). But even a simplistic decomposition can suggest fruitful directions for research and policies to reduce environmentally damaging consumption.

In general, such simple decompositions are variants of the IPAT identity described above. Restating the identity as a "PC" version, in terms of population and consumption:

$$I = P \times C/P \times I/C,$$

where I = Environmental degradation and/or resource depletion; P = the number of people or households; and C = the transformation of energy, materials, and information.

With such an identity as a template and a goal of reducing environmentally degrading and resource depleting impacts, there are at least seven major directions for research and policy. To reduce the level of impacts per unit of consumption, we can *separate* more damaging consumption from more benign forms, we can *substitute* more benign consumption for the more damaging, and we can *shrink* the amounts of environmentally damaging energy and materials per unit of consumption. To reduce consumption per person or household, we can *satiate* some consumption needs and *sublimate* some desires. Finally, we can *slow* population growth and then *stabilize* population numbers as indicated above.

Separate the more damaging from the more benign

Not all of the projected growth in consumption may be resource-depleting – "less available for future use" – or environmentally damaging, that "negatively impacts biophysical systems in such a way as to threaten human health, welfare, or other things people value." Yet, almost any human-induced transformations turn out to be either or both resource-depleting or damaging to some valued environmental component. For example, I recently tracked a year of eight energy-related controversies in Maine. They were related to coal, nuclear, natural gas, hydroelectric, biomass, and wind generating sources, as well as various energy policies. In all the controversies, competing sides, and there were often more than two, emphasized environmental benefits to support their choice and attributed environmental damage to the other alternatives.

Yet despite this complexity, it is possible to rank energy sources by the varied and multiple risks they pose, and for those concerned, to choose which risks they wish to minimize and which they are more willing to accept. We now have almost 30 years of experience with the theory and methods of risk assessment (Cutter 1993, Stern and Fineberg 1996) and 10 years of experience with the identification and setting of environmental priorities (Finkel and Golding 1995). While, there is still no readily accepted methodology for separating out resource-depleting or environmentally damaging consumption from general consumption or to identify those harmful transformations from those that are benign, we can separate consumption into more or less damaging and depleting classes. Taxonomies such as that used to assess the risks of technology (Hohenemser *et al.* 1983) or of environmental hazards (Norberg-Bohm and Clark 1992) can be applied.

Substitute the more benign for the more damaging
Once separated, it is possible to *substitute* less damaging and depleting energy and materials for the more damaging ones. There is growing experience with both encouraging substitution and its difficulties: renewables for non-renewables, toxics with less-toxics, ozone depleting chemicals with more benign substitutes, natural gas for coal, etc.

Perhaps most important in the long run, but possibly least studied, is the potential and value of substituting information for energy and materials. Energy and materials per unit of consumption are going down in part because more and more of our consumption consists of information, writ large, and including much of science and culture. While widely acknowledged, there is relatively little known about such substitutions. For example, do heavily networked organizations travel less than comparative organizations? What would life cycle analysis show for the comparative energy use per unit of value of such information-rich products as music, CDs, or video films?

Shrink energy and material throughputs
Beyond substitution, shrinking the energy and material transformations required per unit of consumption is probably the most effective current means for reducing environmentally damaging consumption. *Stuff: The Secret Lives of Everyday Things* (Ryan and Durning 1997) traces the complex origins, materials, production, and transport of such everyday things as coffee, newspapers, cars and computers and highlights the complexity of reengineering such products and reorganizing their production and distribution.

Yet, there is a growing experience with the 3 "R"s of consumption shrinkage: reduce, recycle, reuse. A growing science, technology, and practice of industrial ecology, combined with the momentum of favorable long-term trends in the efficiency of energy and material usage is encouraging (Frosch 1994, Socolow *et al.* 1994, Vellinga *et al.* 1997). Such a potential, recently led the Intergovernmental Panel on Climate Change to conclude that it was possible, using current best practice technology, to reduce energy use by 30 percent in the short run and 50–60 percent in the long run (Watson *et al.* 1996).

Satiate consumption needs

Addressing the demand-side of consumption, it is possible to reduce our consumption by *satiation* – no more because there is enough. There are of course many signs of satiation for some goods. For example, in the industrialized world, we no longer add refrigerators (except in newly formed households) but only replace them and the quality of refrigerators has improved so that a 20-year or more life is commonplace.

In the financial pages, there are frequent stories of the plight of this industry or that corporation whose markets are saturated and whose products no longer show the annual growth equated with profits and progress. Such enterprises are frequently viewed as failures of marketing or entrepreneurship rather than successes in meeting human needs sufficiently and efficiently. Is it possible to reverse such views, to create a standard of satiation, a satisfaction in a need well met?

Can we have more satisfaction with what we already have, by using it more intensely and having the time to do so (Kempton *et al.* 1995, Merck Family Fund 1995)? Schor (1991) tells of us of some overworked Americans who would willingly exchange time for money, time to spend with family and using what they already have, but are constrained by an employment structure that does not allow for such. Proposed U.S. legislation would permit the trading of overtime for such compensatory time off, a step in this direction.

Sublimate consumption desires

Sublimation is the diversion of energy from an immediate goal to a higher social, moral or aesthetic purpose. Can we have more satisfaction with less, satisfaction derived from the diversion of immediate consumption for the satisfaction of a smaller ecological footprint (Durning 1992, Wackernagel and Ress 1996, Center for a New American Dream 1997)? An emergent research field grapples with how to encourage consumer behavior

---

Box 4.2 Population vs. consumption

· Simpler, easier to study	· More complex
· Well-funded research	· Unfunded, except marketing
· Consensus terms, trends	· Uncertain terms, trends
· Consensus policies	· Threatening policies

---

that will lead to change in environmentally damaging consumption (Jager *et al.* 1997, Vellinga *et al.* 1997).

A small but growing "simplicity" movement tries to fashion new images of "living the good life" (Nearing and Nearing 1990, Elgin 1993). Such movements may never much reduce the burdens of consumption, but they facilitate by example and experiment, other less demanding alternatives. Menzel's (1994) remarkable photo essay of the material goods of some 30 households from around the world is powerful testimony to the great variety and inequality of possessions amidst the existence of alternative life styles.

Can we link such a standard of more is enough to an ethic of enough for all (Institute for Philosophy and Public Policy 1995)? One of the great myths of my childhood, that learning exposed as myth, was that finishing my lunch did not feed the starving children of some far-off place. But increasingly, in sharing the global commons, we flirt with mechanisms that hint at such – a rationing system for the remaining chlorofluorocarbons, trading systems for reducing emissions, rewards for preserving species, or allowances for using available resources.

From more to enough

Let us return to the dialogue with which we began and imagine that in one of the exchanges, each side listened carefully to each other and tried to ask themselves as I have done, just what did they actually know about population and consumption. Struck with the asymmetry that I have described, they might then ask: Why do we know so much more about population than consumption?

My own answer (Box 4.2) is that population is simpler, easier to study, and a consensus exists as to terms, trends, even policies. Consumption is harder, with no consensus as to what it is, and little studies except in the fields of marketing and advertising. But the consensus that exists about population comes from substantial research and study, much of it funded by governments and groups in rich countries, whose asymmetric concern readily identifies the troubling fertility behavior of others and only reluctantly considers their own consumption behavior. So while consumption is harder, we have surely studied it less.

I fear the asymmetry of concern is not very flattering to us in the North and West. We have in our Anglo-Saxon tradition a long history of thought in which the poor are held responsible for their condition – they have too many children – and an even longer tradition of urban civilization feeling besieged by the barbarians at their gates. But whatever the origins of the asymmetry, its persistence does none of us a service. Indeed, the stylized debate of population versus consumption reflects neither popular understanding nor scientific insight. But lurking somewhere beneath the surface concerns is a deeper fear.

For consumption is also more threatening and despite the North–South rhetoric it is threatening to all. In both rich and poor countries alike, making and selling things to each other, including things we might not need, is the essence of our economic system. No longer challenged by socialism, global capitalism seems inherently based on growth – growth of both the number of consumers and their consumption. To study consumption in this light is to risk concluding that a sustainability transition might require profound changes in the making and selling of things and in the opportunities that provides. And to draw such conclusions in the absence of convincing alternative visions is fearful and seemingly to be avoided.

The strategy suggested here is not an alternative vision. But separating out the most serious problems of consumption, shrinking its energy and material throughputs, substituting information for energy and materials, creating a standard for satiation, sublimating the possession of things for that of the global commons, as well as slowing and stabilizing population – all are way-stations on the necessary journey from more to enough.

REFERENCES

Anderberg, S. 1989. A Conventional Wisdom Scenario for Global Population, Energy, and Agriculture 1975–2075. In F. L. Toth, E. Hizsnyik and W. C. Clark, eds., *Scenarios of Socioeconomic Development for Studies of Global Environmental Change: A Critical Review*, RR-89-4, pp. 201–229. Laxenburg, IIASA.

Bender, W. 1997. How Much Food Will We Need in the 21st Century? *Environment* 39(2): 6–11, 27–28.

Bloom, D. 1995. International Public Opinion on the Environment. *Science* 269:354–358

Bongaarts, J. and Bruce, J. 1997. What Can be Done to Address Population Growth?, unpublished background paper for Rockefeller Foundation report *High Hopes: The United States, Global Population and Our Common Future.*

1994. Population Policy Options in the Developing World. *Science* 263:771–776.

Bos, E., Vu., M. T., Masiah, E. and Bulatao, R. A. 1994. *World Population Projections 1994–95 Edition.* Baltimore: Johns Hopkins University Press.

Brown, L. R., Kane, H. and Roodman, D. M. 1994. *Vital Signs 1994: The Trends That Are Shaping Our Future.* New York: W.W. Norton and Co.

Center for a New American Dream. 1997. *Enough!: A Quarterly Report on Consumption, Quality of Life and the Environment.* 1:1 Summer 1997, Published quarterly. Burlington, VT: The Center for a New American Dream.

Cohen, J. 1995. *How Many People Can the Earth Support?* New York: Norton.

Cutter, S. L. 1993. *Living with Risk.* London: Edward Arnold.

Davis, K. 1990. Population and Resources: Fact and Interpretation. In K. Davis and M.S. Bernstam eds., *Resources, Environment and Population: Present knowledge, Future Options*, pp. 1–21. Supplement to Vol. 16, 1990, *Population and Development Review.* New York: Oxford University Press.

Deevey, E. 1960. The Human Population. *Scientific American* 203:194–204.

Dietz, T. and Rosa, E. 1994. Rethinking the Environmental Impacts of Population, Affluence, and Technology. *Human Ecology Review* 1:277–300.

Durning, A. 1992. *How Much is Enough?* New York: Norton.

Elgin, D. 1993. *Voluntary Simplicity; Toward a Way of Life that is Outwardly Simple, Inwardly Rich.* New York: William Morrow.

Finkel, A. and Golding, D., eds. 1995. *Worst Things First: The Debate Over Risk-Based National Environmental Priorities.* Washington, DC: Resources for the Future.

Frosch, R. 1994. Industrial Ecology: Minimizing the Impact of Industrial Waste. *Physics Today* 47(11):63–68.

Gleick, P., ed. 1993. *Water in Crisis: A Guide to the World's Fresh Water Resources.* Oxford: Oxford University Press.

Hohenemser, C., Kates, R.W. and Slovic, P. 1983. The Nature of Technological Hazard. *Science* 220: 378–384.

Institute for Philosophy and Public Policy. 1995. The Ethics of Consumption. *Philosophy and Public Policy*, Report from the Institute for Philosophy and Public Policy, School of Public Affairs, University of Maryland, Baltimore, 15(4).

Jager, W., van Asselt, M., Rotmans, J., Vlek, C. and Costerman Boodt, P. 1997. *Consumer Behavior: A Modeling Perspective in the Context of Integrated As*sessment of Global Change. RIVM Report No. 461502017. Bilthoven: National Institute for Public Health and the Environment.

Kane, H. 1997. Half of Languages Becoming Extinct. In L. Starke, ed., *Vital Signs 1997*, pp. 130–131. New York: Norton.

Kates, R. 1994. Sustaining Life on the Earth. *Scientific American* 271: 116–123.

1995. Labnotes from the Jeremiah Experiment: Hope for a Sustainable Transition. *Annals of the Association of American Geographers*, 85(4): 623–640.

1997. Population, Technology, and the Human Environment: A Thread Through Time. In J. Ausubel and H. Langford, eds., *Technological Trajectories and the Human Environment*, pp. 33–55. Washington, DC: National Academy Press.

Kempton, W., Boster, J.S. and Hartley, J.A. 1995. *Environmental Values in American Culture.* Cambridge, Massachusetts: MIT Press.

Keyfitz, N. 1981. The Limits Of Population Forecasting. *Population and Development Review* 7:579–593.

Knodel, J. and van de Walle, E. 1976. Lessons from the Past: Policy Implications of Historical Fertility Studies. *Population and Development Review* 5:217–245.

Lee, R. 1989. Long-run Global Population Forecasts: A Critical Appraisal. In K. Davis and M.S. Bernstam eds., *Resources, Environment and Population: Present knowledge, Future Options*, pp. 44–71. Supplement to Vol. 16, 1990, *Population and Development Review*. New York: Oxford University Press.

McKewon, T. 1976. *The Modern Rise of Population*. New York: Academic Press.

McNicoll, G. 1992. The United Nations' Long-Range Population Projections. *Population and Development Review* 18:333–340.

Menzel, P. 1994. *Material World: A Global Family Portrait*. San Francisco: Sierra Club Books.

Merck Family Fund. 1995. *Yearning for Balance: Views of Americans on Consumption, Materialism, and the Environment*. Takoma Park, MD: Merck Family Fund.

Myers, N. 1997a. Consumption in Relation to Population, Environment, and Development. *The Environmentalist* 17:33–44.

1997b. Consumption: Challenge to Sustainable Development. *Science* 276:53–57.

Nakicenovic, N. 1996. An Energy Primer. In R. Watson, M. Zinyowera, and R. Moss, eds., *Climate Change 1995: Impacts, Adaptations and Mitigation of Climate Change: Scientific-Technical Analyses*, pp. 75–92. Cambridge: Cambridge University Press.

Nakicenovic, N. 1997. Freeing Energy from Carbon. In J.H. Ausubel and H.D. Langford, eds., *Technological Trajectories and The Human Environment*, pp. 74–88. Washington, DC: National Academy Press.

Nearing, H. and Nearing, S. 1990. *The Good Life: Helen and Scott Nearing's Sixty Years of Self-Sufficient Living*. New York: Schocken.

Norberg-Bohm, V. and Clark, W.C. 1992. *International Comparisons of Environmental Hazards*. Center for Science and International Affairs Discussion Paper 92–09. Cambridge, MA: John F. Kennedy School Of Government, Harvard University.

Pimm, S., Russell, G., Gittelman, J. and Brooks, T. 1995. The Future of Biodiversity. *Science* 269:347–350.

Population Reference Bureau. 1998. *1998 World Population Data Sheet of the Population Reference Bureau*. Washington, DC: Population Reference Bureau.

Raskin, P, Chadwick, M., Jackson, T. and Leach, G. 1995. *The Sustainability Transition: Beyond Conventional Development*. Boston: Stockholm Environmental Institute.

Royal Society of London and the U.S. National Academy of Sciences 1997. Towards Sustainable Consumption. Reprinted in *Population and Development Review*, 23(3): 683–686.

Ryan, J. and Durning, A. 1997. *Stuff: The Secret Lives of Everyday Things*. Seattle, WA.: Northwest Environment Watch.

Schipper, L., Ting, M., Krusshch, M., Unander, F., Monahan, P. and Golove, W. 1997. *The Evolution of Carbon-Dioxide emissions from Energy Use in Industrialized Countries: An End Use Analysis.*, LBL-38574. Berkeley, California: Lawrence Berkeley National Laboratory.

Schor, J. 1991. *The Overworked American*. New York: Basic Books.

Socolow, R., Andrews, C., Berkhout, F. and Thomas, V., eds. 1994. *Industrial Ecology and Global Change*. Cambridge: Cambridge University Press.

Stern, P. and Fineberg, H., eds. 1996. *Understanding Risk: Informing Decisions in a Democratic Society*. Washington, DC: National Academy Press.

1997. Toward a Working Definition of Consumption for Environmental Research and Policy. In Stern, P., Dietz, T., Rutaan, V., Socolow, R.H. and Sweeney, J.L., eds., *Environmentally Significant Consumption: Research Directions*, pp. 12–25. Washington, DC: National Academy Press.

Stern, P., Dietz, T., Rutaan, V., Socolow, R.H. and Sweeney, J.L. eds. 1997. *Environmentally Significant Consumption: Research Directions*. Washington, DC: National Academy Press.

Stoto, M. 1983. The accuracy of population projections. *Journal of the American Statistical Association* 78: 381:13–20.

Telegeography 1997. http://www.telegeography.com/Publications/tg96_intro.html (6/12/98).

United Nations, Department of International Economic and Social Affairs. 1992. *Long-Range World Population Projections: Two Centuries of Population Growth 1950–2150*. New York: United Nations.

United Nations, Population Division of the Department of Economic and Social Affairs at the United Nations Secretariat, 1998. *World Population Projections to 2150*. New York: United Nations.

Vellinga, P., de Bryn, S., Heintz, R. and Mulder P., eds. 1997. *Industrial Transformation: An Inventory of Research*. IHDP-IT No. 8. Amsterdam: Institute for Environmental Studies.

Vincent, J. and Panayotou, T. 1997. . . . or Distraction? *Science* 276:53–57.

Wackernagel, M. and Ress, W. 1996. *Our Ecological Footprint: Reducing Human Impact on the Earth*. Philadelphia, Pennsylvania: New Society Publishers.

Walter, J. and Schofield, R. 1989. Famine, Disease and Crisis Mortality in Early Modern Society. In J. Walter and R. Schofield, eds., *Famine, Disease and the Social Order in Early Modern Society*, pp. 1–73. Cambridge, UK: Cambridge University Press.

Watson, R., Zinyowera, M., Moss, R., eds. 1996. *Climate Change 1995: Impacts, Adaptations and Mitigation of Climate Change: Scientific-Technical Analyses*. Cambridge: Cambridge University Press.

Wernick, I. 1996. Consuming Materials: The American Way. *Technological Forecasting and Social Change* 53: 111–122.

Wernick, I. and Ausubel, J. 1995. National materials flow and the environment. *Annual Review of Energy and Environment* 20:463–492.

Wernick, I., Herman, R., Govind, S. and Ausubel, J. 1997. Materialization and Dematerialization: Measures and Trends. In Ausubel, J. and Langford, H., eds., *Technological Trajectories and the Human Environment*, pp. 135–156. Washington, DC: National Academy Press.

Wernick, I., Waggoner, P. and Ausubel, J. 1998. Searching for Leverage to Conserve Forests: The Industrial Ecology of Wood Products in the U.S. *Journal of Industrial Ecology* 1(3):125–145.

World Resources Institute, United Nations Environment Programme, United Nations Development Programme and World Bank. 1996. *World Resources, 1996–97*. New York: Oxford University Press.

Worldwatch Institute. 1998. *Database Disk*, Spreadsheet TOTENRG.WK1. Washington, DC: Worldwatch Institute.

5
—————

Spirituality and sustainability

On all sides, the future and purpose of our industrial society is being called into question. This comes somewhat as a surprise. Not so many years ago, we were told that we had never had it so good, and that in time, things would be even better, as we – the developed nations – with the "underdeveloped nations" in tow, would march into the Age of Plenty. Modern science and technology had done it; Western civilization had done it.

It was a good dream.

I came of age in the 1960s. I came into a sense of self and world and future in those years, the defining moment of which, for me, was the day the men landed on the moon.

I was at summer camp. We were sitting on the floor of a lean-to at Pioneer Village, into which a black and white TV had been wired, then rolled into the hall for the special occasion, and sitting in our T-shirts and shorts, 150 eager campers stared, moonstruck, up at the screen. What a sense of collective triumph and pride and optimism, and – invincibility. I came of age also as the daughter of an immigrant from Eastern Europe, who landed in New York City circa 1945. And so I came up knowing something about the promise of the American Dream, and the hopes it unwaveringly carried for its children. By so many, it was intended as, and assumed to be – a good dream.

The problem with the dream, of course, and its promises is that we face limits, limits on this planet's carrying capacity for our human numbers, limits on using nature as the source of our food, fuel, minerals and the dumping ground for our wastes. And limits too, we must say, on what we can expect from technological innovation. We are, after all, human and not God.

The faith community is not uncomfortable with the notion of limits. In fact, it's our thing. The notion of limits is at the heart of who we are, and oddly, the hope we proclaim as Church. Nineteen days ago was Ash Wednesday, the opening of Lent, the penitential season of the Christian community. On that day, upon many a forehead ashes were placed with the sign of the cross, and the words, "Remember that you are dust and to dust you shall return". These words are not merely a reminder of the inevitability of our individual deaths, but an invitation to spiritual renewal, *metanoia* which – as all growth – requires change and suffering, even dying, in order for something new to be born.

With a Lenten sensibility in mind, the Church proposes that the environmental crisis is one principally of the heart and of the will, demanding conversion (turning around) at the deepest level of our individual and communal beings. The theologian Larry Rasmussen (Union Theological Seminary, New York City) attests that those of us in the developed world have a major moral dilemma, which is that "neither the human family nor the rest of nature can afford the modern world; and yet we cannot extract ourselves from it and achieve a different order without a period of wrenching, costly change". And so in what follows I will attempt to speak of both the cost and the promise of living in a new moral order, a shift which will demand costly change, but which will greatly benefit not only other human beings but all of nature.

What is basic is transformation of our world view, a perception of our place and role within the natural world, changing from a role of conquest and profit to one of interdependence and reverence. One of the most obvious centers for changing the way people perceive the world and how they relate to it is the religious community (church, synagogue, meditation circle). By their very nature local, these communities can be places of support, inspiration, modeling and ultimately, social change. The role of church communities during the early years of the movement for civil rights for African Americans is only one example of parish communities playing just such a role.

I am not unaware that I am the only representative of the religious community in this volume. I am confident, however, that some of you worship in a church, synagogue or mosque and I am equally confident that for many others of you, your work in sustainability is at heart a spiritual issue, drawing upon both your deepest moral convictions. In either case, I say to us all, "Be transformed!" Be transformed, as the apostle

Paul wrote in his letter to the Romans. Nothing short of our moral and spiritual transformation can accomplish the tasks before us in these crucial years ahead.

Idolatry and the American dream

In 1933, Aldo Leopold, author of *The Sand County Almanac*, and the man who coined the concept of the "land ethic", wrote in an unpublished essay entitled "The Ecological Conscience":

> We can be ethical only in relation to something we can see, feel, understand and love ... No important change in ethics was ever accomplished without an internal change in our intellectual emphasis, loyalties, and convictions. The proof that conservation has not yet touched these foundations of conduct lies in the fact that philosophy and religion have not yet heard of it.

Leopold's challenge to the religious institutions of our civilization has at last been heeded most demonstrably in the last 10 years, although important foundations were laid in the 1970s. The Church is beginning to emphatically assert the first article of its principal credal statement, the Nicene Creed, "We believe in God, Creator of heaven and earth" and interpret this confession of faith in God as creator in light of the environmental crossroads at which our civilization stands.

Although the Church as institution in any broad and public way is relatively a 'Johnny come lately' to the sustainability conversation, the Church nonetheless brings some unique and vital strengths to the work ahead. I would like to call to your particular attention the notion of idolatry. The Church understands that all human visions and constructs are incomplete and fallible; therefore, there is always need for re-formation, re-formulation of even the most compelling of dreams. The so-called "American dream" of prosperity for all, based as it is on the unqualified goodness of infinite economic growth – itself based on the assumption of infinite natural resources – is such a vision.

The Church grounds its thinking on idolatry on the first article of the Decalogue: "Hear the commandment of God to his people: I am the Lord your God. You shall have no other gods but me. You shall not make for yourselves any idol". As I will suggest further on, the Church is uniquely positioned to challenge the prevailing world view that views earth as commodity, a values system that theologian John Cobb and economist Herman Daly call "economism".

Allow me to clarify the Biblical understanding of idolatry. Briefly stated, idolatry is the human propensity to make an ultimate of things that are not ultimate: the enshrinement of any other person or thing in the place of God, to put at the center things that may be true but are not ultimately true. Thus human beings as idol makers have from time to time worshipped stones and suns and also monoliths of themselves or their progenitors: we have worshipped many Caesars. That is part of what it is to be human.

The dream of the post-World War II industrial world has reached its nadir. This dream has been based on many erroneous assumptions, principal of which is that the natural "resources" of the earth will last forever. This sacred cow of infinite progress, consumption and development, we now realize, is based in an idolatrous vision that needs to be challenged and ultimately replaced. The task of sustainability is not just restoration of a nature despoiled, nor is it to simply slow down the damage, so that it becomes "sustainable", although these tasks are all essential: but we must concurrently change the way we think, feel and love, and encourage one another to do so, and then the right actions have a chance to follow. In short, the ecological crisis itself is not first of all ecological – it is a crisis in the way we see ourselves and the world.

This means that in order to deal effectively with the vast web of ecological problems we have to change our world-image, and in turn this means that we have to change our self-image. This is the thing about which we must be absolutely clear if we are to find a way out of the hell to which we have condemned ourselves.

"Who am I?" is a question that the great thinkers of every age and every region have asked, among them Lao Tzu, Aristotle, Descartes, and Kant. Since the Industrial Revolution, this question has been implicitly answered in the West, by defining human as consumer. Thus "economism" has been the implicit religion of Western culture.

The idolatry of this definition must be challenged, for it impels us to look upon ourselves as little more than bipeds whose destiny and needs can best be fulfilled through the pursuit of social, political, and economic self-interest. An illustration is telling. Bill McKibben, author of *The End of Nature*, recorded 2400 hours of videotape from one day's worth of cable TV, then spent one year analyzing it. He concluded that the message came down to a single notion being put across: "You, the viewer, are the most important thing on the face of the earth. Your immediate desires are all that count. We can satisfy them. We do it your way. This Bud's for you!" (McKibben 1990). We might call this image of the self "homo consumens", that is, the human whose very being is encapsulated as one who buys products.

Hence we find ourselves living in a world that sees nature as 'environment', a soulless resource for food, wealth and recreation, all of which we think we are to entitled to exploit by any technique we can devise.

Thus, having in our own minds de-sanctified ourselves, we have de-sanctified nature as well. This twisted (sinful) self-image and worldview have their origin in a loss of memory. We have forgotten who we are and who God is, and the nature of life as gift. So long as we persist in this course, if present trends continue, we won't. Our idolatry of Economism shrinks our understanding of the human condition, undermines community and threatens the health of the earth upon which all creation utterly depends.

Panentheism

Having suggested something of an anthropology, let me focus briefly on sacred cosmology, a notion with which I suspect many of you are familiar, if only intuitively. "The world is charged with the grandeur of God" (Gerard Manley Hopkins 1953). There is a sacredness about the earth that all world religions, most notably those of indigenous peoples, have always recognized. For the Christian, the earth is not ' environment', a value-neutral word but creation, that is, created, blessed and sustained by the Spirit of God, bearing the stamp of the Creator. Thus nature is profoundly relational. Some examples from traditional Western Christian thinkers may help illuminate this further.

For Augustine, the natural world was in fact a vehicle for the Spirit's self-revelation: "Sacred writings are bound in two volumes – that of Creation and that of the Holy Scripture". His thinking became foundational for what was to follow in the tradition of Natural Theology.

For St. Thomas Aquinas: "Because God in his goodness could not adequately be represented by one creature alone, he produced many and divers creatures, . . . and hence the whole universe together participates in the Divine goodness more perfectly and manifests it better than any single creature whatever". This may have been the first argument for bio-logical diversity!

For St. Bonaventure: "Every creature can be understood as the self-expression of the Word of God".

Finally, perhaps in the fullest manner, the mystical and contemplative traditions of the Church, and many of the monastic writings and tradi-tions, embody a deep spirituality of creation that constituted a formid-able, if not consistent, challenge to the values of their time, and can serve as a theological foundation for our own. To summarize, there is moral worth in all that is, beyond its utilitarian value for humans. Nature is worth saving because it is sacred.

The most convincing case, for example, for saving from extinction the resplendent quetzal, a glittering golden-green and ruby bird in Central America – is simply that the loss of such a bird, like the destruction of a Mozart sonata or a Renoir painting or Saint Sofia in Istanbul, would be a loss of something beautiful and unrepeatable. It is sacred gift.

There are strands of this sacred cosmology throughout Christian Scripture, tradition, and hymnody. It is time that the Church and other world spiritual traditions begin to recognize, assemble, and plumb the depths of this spirituality from our respective traditions. In fact, over the last five years or so, I have been encouraged to see, and can with pleasure bear witness to the fact that there is in fact a significant religious renewal in our time – understanding nature as one of the realms of the Divine. Perhaps we are on the cusp of the Third Great Awakening.

Ecojustice

I want now to speak of something known widely as ecojustice or environ-mental justice, that is, an environmental world view that is intimately connected with human justice, especially for the poor. This understand-ing underlines the increasingly well known fact that environmental deg-radation is economically specific: The poor suffer much more than the well-to-do. This is true in North America, but even more so when one compares the relatively small populations of the industrialized West with those of the developing nations.

I will begin with a story. In 1991, at the first meeting of the Presiding Bishop's Environmental Stewardship Team, (the first national committee of the Episcopal Church in the USA concerned with environmental matters) I gathered with my fellow team members as one representing the

Pacific Northwest. At our first liturgy together, the representative from Costa Rica, a priest, said something in his homily which I have never forgotten. After reflecting on the reading from Paul's letter to the Romans, he spoke of the environmental movement in North America, and he concluded with this riveting question: "You North Americans, when you speak so passionately of saving the earth for the grandchildren, for whose grandchildren are you saving the earth?". I have yet to hear such a powerful and poignant summary of ecojustice.

The Church is uniquely positioned to insist that care of the earth and care of the human – especially the poor who are most severely impacted by environmental destruction – can never be separated. I come to you from the land and waters of the spotted owl and the salmon, from communities in the Pacific Northwest where it is abundantly clear that the separation and polarization of human need and nature's need simply will not work. It has been, in fact, highly counterproductive.

The Church is also in an important position to enter into the conversation around population and the concerns regarding the soaring numbers of our kind expected in the next century. Here the Church is learning that there are two kinds of population pressures on the earth: first, the sheer number of people living at subsistence level, and second, the overconsumption of the developed world's smaller numbers of people (the average North American consuming annually approximately 30 times the amount of natural resources than our counterpart in the developing nations). Both of these factors constitute serious population pressures on the earth.

It is significant to point out here that the Church is learning from the many mistakes made in the heyday of its missionary zeal. In those years it was normative to impose cultural assumptions and cultural values on other cultures in the name of religious fidelity. This was often a destructive and not a life-giving act. Today, just as missionary activity is now recognizing the validity of the inherent spirituality of the country where it is, we must also recognize the unique concerns and perspectives around population issues that are endemic to these cultures. The developing nations, as was loudly stated at the Earth Summit in Rio in 1992, do not want to hear accusations from the developed world about population unless we are willing to recognize both our own levels of consumption and the complex factors involved in their rising numbers.

In short, the Church is learning that we in the developed nations must come to the table not as the folks with the answers to complicated cultural, religious, and environmental issues, but as people ready to engage with full respect in these crucial discussions on population, knowing that the solutions must be joint, profoundly intercultural solutions.

The Church has had a long history of focusing on the needs of the poor and dispossessed, and has had some saints, notably Francis of Assisi, whose love of both the human poor and of the natural world were inseparable. He loved both. He served both. He refused unholy separation or choosing. And so, when the Church belatedly entered into the environmental conversation it brought with it an awakening of the issues of eco-justice to the environmental community. It is of interest that in recent years, many members of the environmental community acknowledge that justice for the human poor has been a dimension of environmental change that has frequently been ignored or minimized, a practice which is being critically re-evaluated.

From economism to earthism

I have been speaking of changing the way we see ourselves and the way we see the world. I have been reminding us of the inseparability of the needs of the natural community, plant, animal, and human. I have brought to mind the vision of the world as sacred gift which is found in our religious traditions. I have injected the theological concept of idolatry into our conversation, the sense that God is absolute but no one human ideal is absolute.

I now want to say a few more things on the topic of idolatry. Broadly defined, this is the way the Bible sees the problem: We have a choice to love the Lord our God with all our heart and mind and soul and strength, or not. We are free to acknowledge the world as God's creation or not. The recognition that all things derive from God at every moment and are in utter dependence on God, furthermore the confession that none of this had to be, the acknowledgement and the acceptance of this in thanksgiving and humility is called faith. The refusal to acknowledge the world's origination and destination in God and the attempt to make the world a self-sufficient reality is called sin.

It follows that while God is absolute, no one idea of God or of the "ultimate good" or world-view is absolute. One of the seminal thinkers in the Church today is John Cobb, who as previously mentioned collaborated with economist Herman Daly in the work *For the Common Good*. Cobb coins the word 'economism' to suggest that, since World War II, the vision of the world that has dominated our collective consciousness and values is economic growth. So much has this vision dominated our thinking and acting, so much has it subsumed other ways of thinking that economic growth has in fact become the operative religion of our times. According to Cobb, what shapes the value system of a culture is its operative religion.

From the perspective of biblical faith, all religion – in fact particularly religion – is prone to idolatry, for when religion is not in service to a larger good, it becomes itself the good under which all other goods are subsumed. This is what has happened to the proximate good of 'economic growth'.

What Cobb and other theologians counter is that the economy as we understand and measure it, needs to be in sync with 'the Great Economy', that of the whole family, humanity and nature, of which we are a part. Global economy needs to be in sync with earth's economy or it will literally go out of business.

So in theological terms, economy defined as economic growth has become bloated into 'economism' and this is idolatrous. Moving from this model to 'earthism' is a healthy corrective, and a response that opens us as human beings to say "yes" to God's claims on the world, to say "yes" to the frightening contingency of human existence and the planet's existence. None of this had to be here. And God is holding it up at every moment. As gift. What follows is a manner of living that takes seriously the claims for health and longevity of the whole creation.

The church as a center for empowerment of local communities

I will suggest here that the church, specifically the local community church, is the laboratory and the crucible in the community in which we return to our senses. Please be clear that I also understand the synagogue, the mosque, etc., in the same way in this regard.

"This land is your land, this land is my land", so sang Woody Guthrie, troubadour of our country. This song is an American hymn of sorts, although in saying this I recognize the supreme arrogance of these opening lines, in light of the millennia in which native peoples have lived on this continent and called it home! In tent revival style, we sing it, and it reminds us that the future of transforming the world begins in our own backyards and local communities. Local religious communities are ideal places for members and non-members alike to get together and talk, asking each other some good questions, such as, "Is there anything

better than the good life? . . . How much is enough? . . . What can I do? . . ."
Church basements are good places to begin to understand the ecological
consequences of our actions as a community and to begin to strategize
alternatives.

Parish churches are unique in many communities for the diversity of
political opinions and economic classes from which people come.
Churches are one of the rare places where people of differing perspectives
and opinions on environmental issues come on a regular basis to pray and
work together. Might not the church be a place in the society where some
of the chasms of opinion and stereotype might be bridged?

In short, the local church is one of the important centers of bioregion-
alism: the parish church is in and a part of the local economy. Potentially,
local churches can be counter-cultures in the wider culture that challenge
the status quo (that local communities exist for the sake of business's
profits). No, business and government exist for the sake of local communities.

Once again, I can bear witness that this is happening. Parish churches
across the country and across the denominations are in fact beginning to
awaken to environmental issues as spiritual and moral issues for our
time. Change is happening.

Spiritual and moral transformation

I began with a quote from Aldo Leopold, having to do with moral values
being capable of transformation only with regard to what a person can
see, know and love, and with an imperative from the Apostle Paul – "Be
transformed". This may be seen as an unusual juxtaposition of texts;
however both Paul and Leopold pose a challenge to the assumptions and
marching orders of Western civilization. Neither morality nor cultural
change can be legislated. Nor can it be simply taught in a curriculum. In a
recent journal on environmental ethics was the sentence: "Education
isn't the solution because ignorance isn't the problem". Now clearly when
it comes to environmental sustainability in the years ahead, both
reformed and enlightened legislation and education alike are crucial.
However, from the religious perspective, these strategies are not enough,
and in themselves are inadequate to the task that lies ahead.

One has to be moved to change, motivated consistently to make the
shifts and the sacrifices needed. This movement can come only from the
heart. Leopold tells the story of his own transformation, how his view of

the world and of himself changed. Shooting into a pack of wolves in then-rural Wisconsin, he watched "the green fire die" in the eyes of the she-wolf. It was her death that brought a part of Aldo to life and was to influence his life's work. Roger Tory Peterson (author of *Field Guide to the Birds*), recalling the experience that changed his life and set him on the course he was to follow, called it a "trigger". On a walk he took as a junior high student one Saturday with a friend, he came upon a flicker in an oak tree. Thinking the bird was dead, he poked at it gingerly, but the bird was just asleep, probably resting from migration. When he touched it, its eyes sprang open and it flew away. "This inert bunch of feathers suddenly sprang to life." What struck him was the contrast between what he thought was dead, but was very much alive. "Almost like resurrection . . . Ever since then birds seemed to me the most vivid expression of life." Such changes of heart can leave their mark not only on the individual life, but on the life of the culture as well.

The Church might rightly be asked: has there been another era in the human story in which a major shift in values has occurred? It seems to me it can be argued that in the last two centuries alone this has in fact occurred: the Emancipation Proclamation and in the decades which followed, the Civil Rights movement, the Women's Suffrage movement, the unionizing movement early in this century, and finally the Endangered Species Act. In all of these 'movements', there was in fact significant and counter-cultural movement in ethical perspective: steps in human understanding were taken that substantially widened the circle of ethical consideration. The ethical norm was no longer 'me and my family' or even, as in the case of the Declaration of Independence "all men are created equal". In each of these movements the circle of moral worth expanded outward – from "all white men who were free and property owners" (Declaration of Independence) to people of color, women, working people of modest means, and other species.

We can therefore look with hope, which is not the same as optimism, as we look back on our history, in this nation alone. There have in fact been significant shifts in our understanding of who is worthy to be and to flourish in our world.

"Be transformed." That is, come home to yourself and to your God and to the place you have been gifted to live. Hold your fears for the world and its future lightly, knowing that you are not alone. Look for surprising grace that is at work in the world. Carry high the torch of hope that you have. It's catching.

REFERENCES

Hopkins, G. M. 1953. *Gerard Manley Hopkins: Poems and Prose*, p. 27. Baltimore: Penguin Books.

Leopold, A. 1933. *A Sand County Almanac*. New York, NY. Oxford University Press.

McKibben, B. 1990. *The End of Nature*. New York: Doubleday.

6
―――――――――

Leadership skills for sustainable development

Preamble on personal restraint

Twenty-five years ago this month, while still an undergraduate candidate at Cornell, I came across a passage that convinced me I could become a writer of books someday. It was written by Marguerite Wildenhain, the Bauhaus potter in her superbly written book *The Invisible Core*, and it reads with a feeling of moral obligation even today, as follows:

> There is no reason to be proud of whatever gifts one has . . . But there is a reason to be deeply thankful for them . . . The more capacity a person has, the greater and more cogent will be the moral obligation to do something honorable with what he has. (WILDENHAIN, 1973)

The passage does not end here, with a cliché, and elitist sense of noble obligations, but instead, returns to a more earthy and sincere point about any craft person's devotion to their way of life:

> Let us look at the implication, man like animals, is by nature lazy, but the creative man would always work. He works not only because he unconsciously acknowledges his ability to do so with the acceptance of deeper human responsibility. He understands that his work-time cannot be what it is for most men, from a certain hour to another definite hour. His work will be his total life, no less. *That* he feels to be the least he can do to make up for the gift of abilities that he was given.
> (WILDENHAIN, 1973)

For those that have admired Wildenhain's pots, and have paused before their astounding functional beauty, you know she lived this kind of devoted life. Despite tremendous pressures from the marketplace to increase the output of her work, she resisted these external pressures in a deliberate effort to capture "the invisible core" of each pot. Her passage ends with a point I now find strangely relevant to the current task of reflecting on leadership skills for sustainable development, since so few leaders in history have this kind of artistic restraint she demands of herself:

> Needless to say, there are innumerable temptations along the path of a craftsman to do things badly, to evade responsibility and effort, to look at life, flippantly, and one succumbs to these often enough. As a whole, though, this concept of total daily and complete devotion to his work becomes for the mature craftsman not only his moral law, but his deepest concern and his most constant joy . . . this artistic and ethical pressure that permeates his whole life, in work and leisure, in joy and sorrow, will give his work the valid human quality that is the sign of the work of art and out of a technician a creative artist will have developed.   (WILDENHAIN, 1973)

What follows is an essay designed to explain one claim: While most popular management literature can describe how leaders spend their time, staff, and resources on making sure their organizations have a promising future, this same literature is essentially bankrupt regarding the topics of personal restraint and professional devotion evident in the Marguerite Wildenhain passage. Yet deep in our hearts most of us know that sustainable development is not achievable without leaders who can teach the values of such restraint, quality, and devotion on a global scale.[1]

Across the last 25 years I have found similar passages in the admired works of writers as different as William Hazlett, Henry David Thoreau, and George Orwell, and as I suspect now that the "leadership principles of strategic restraint" lies buried in the many diverse works of the world's greatest spiritual literature.

Yet to keep this essay simple and readable, please return to Wildenhain's passage as an example of what is often missing in the post-World War II debates about sustainability. I seldom see any serious examination of the role of personal restraint in the political, scientific, or corporate claims about sustainable enterprises. While Schneider, Kates, and Daly refer to the inevitable need of altering our rates of consumption and the scientific and economic infrastructure promoting such robust rates of consumption, these three authors' thoughtful commentary are but small whispers in the clanging echoes of the extant literature on the subject of our future.

Of course, at this point, most rationalists may ask: Is there any room for restraint in our new paradise of global consumers? Yet what gives us some wiggle room to outsmart the remorseless logic of rampant consumerism is the fact that throughout history leaders can preach about the powers of restraint in credible ways. Here one need only think of Winston Churchill's famous appeal about blood, sweat, and tears, or remember Lincoln at Gettysburg or Caesar explaining the subtle necessities of Roman sacrifice. Throughout time, leaders emerge as individuals, who discover in us the emotional intelligence it takes to get a job done, even when that set of tasks requires sacrifices and sustained restraint.

Leaders for sustainable development will, across time, define a new business logic. While the dominant language and metrics of that logic must include refinements of the conventional corporate strategic concerns about time, quality, cost reduction, distribution, and critical staffing issues, one can begin to see the emergence of these new leaders in the realm of corporate environmental management. I first sensed this while writing my 1990 book, with Peter Asmus, for Simon and Schuster, *In Search of Environmental Excellence: Moving Beyond Blame*. In that book, I assembled a short chart to distinguish some new emerging aspects of environmental leadership. What follows presents an end-of-the-century update of that chart (Table 6.1).

Across my consulting practice I have seen hiring managers select their staff based on the conserving and restraint-based values seen on the chart's right column. Admittedly, this is a rare event. But it does happen more and more, especially because of downsizing.

As we all well know, the longing for success built into most business training has little to do with such strategic conservation of resources. Moreover, the supreme machinery of marketing can't directly ask the important key questions about "what is enough." Since World War II, most leaders needed to center their efforts along one line of industrial thinking, and this strong bold line allowed expansive celebration of three kinds, or areas, of innovation:

1. New product development
2. Rights of consumers
3. Transfer of technologies.

Government, civil discourse, and corporate policy have, since World War II, essentially centered their agendas along these lines. The result is a well-advertised net increase in the world's rate of consumption, especially in America and other advanced industrial nations.

There have been some significant countervailing developments.[2] Take, for example, our new options of e-mail over express mail, or some

Table 6.1. *Next century's fork in the road*

I. Adversarial politics – business as usual	I. Beyond blame strategies

· End-pipe regulations
· Legal litigation for problem solving
 	
· Enforcement and fines as noted
  performance measures
	· Solution-oriented public strategies
	· Building from a common ground of
	  science and law
	· Education and development, not
	  just technology and products

II. Consumer society	II. Less is more approach to growth

· Inability to distinguish between
  wants and needs
· Materialism: the American dream
  held as the key model
· Reliance on the superabundance and
  resilience of nature
	· Self-sufficiency and minimalism in
	  organizations and profit centers
	· Conservation values towards natural
	  resources in day-to-day actions
	· Environmental and resource
	  management are part of corporate
	  strategy

III. Dominant culture dynamics	III. Emergent culture dynamics

· Nationalism's inability to act as key
  to lobbying
· Creator/destroyer attitude dominant
 	
· Short-term usage valued
	· Global perspective; grassroots
	  activism – both seen as related
	· Living as an implicit part of nature,
	  not above nature
	· Long-term planning as valued

Source: Updated by Bruce Piasecki for AHCGroup, first published in *In Search of Environmental Excellence* (Simon & Schuster, 1990)

select national refusals to follow the American expansionist economic model blindly. Yet despite each small step forward, we sense the rapids of industrial development wherever one looks. As a result, most of my students think of sustainable development as a joke, or at best, as an oxymoron (Cairncross honestly characterizes sustainable development as an ideal in her smart, and cautious contribution to this book).

Nevertheless, many smart people today wonder if the next generation of environmental leaders can apply the rigor and discipline of Wildenhain's art to their corporations. While it is too early to judge how much of this intent is grounded in promotional appeal only, the next section explores this new interest of leaders at length.

What is environmental leadership?

Leadership is not something that is done to people, like fixing their teeth. Rather, it is what unlocks their people's potential, challenges them to become better, calls them to task for the lies they have told themselves. (Bill Bradley)

Have you ever noticed how seldom environmental leadership is defined clearly and intelligently? Business books abound with cases that explore what creates a leader within a private-sector organization. Environmental books, on the other hand, have emphasized wrongdoing, imminent doom, and identification and solution of technical problems.

Seldom do these two traditions meet. Yet with the increasing importance being attached to the qualities of leadership, it is time to look at what distinguishes environmental leadership from other kinds of superior corporate performance.

Environmental leaders in the corporate arena face a challenging set of demands that differ from those faced by their corporate peers in other, more defined and established departments such as finance, sales, and marketing. First, they must achieve regulatory compliance. Second, they must go beyond compliance to recognize business opportunities while being able to take on prudent business risks. Third, they must work skillfully with a wide range of external stakeholders, not all of them friendly. Environmental leaders, then, require an extraordinary range of knowledge, diplomatic and political talent, dispute-resolution abilities, basic business skills, and a humanism in their decision-making that reaches beyond this quarter's balance sheet. It is this range of know-how that equips such professional managers to offer significant insight into the debates over the next 50 years, and about the difficult choices regarding resource use that must be made.

The question then arises: What skills does the individual manager need to excel in the environmental arena? Although many of the skills they need are shared by political or corporate leaders, what's unique for environmental path-breakers is the general comprehensiveness of their skills. A second special feature is how often environmental leaders must initiate and guide company-wide change with limited staff and resources. This should prove useful training to help organizations live up to their claims about sustainable development.

Leadership goals and sustainable enterprises

The first goal of environmental leaders is to achieve compliance cost-effectively. Much has been written about the complexity of compliance, from the overlapping and often conflicting levels of government to the sheer volume of data that must be managed. To reach this goal, the leader must comprehend legal, engineering, and scientific needs, and also make the goal understandable to others within the firm, especially the chief executive officer and product champions.

Yet compliance is merely the starting-point, not the finish line. Leaders also pursue a second goal: achieving compliance without extinguishing the spark for risk-taking, innovation, and business advances. Here one achieves a productive balance between regulatory demands and business expectations.

The third goal of environmental leaders is to answer public expectations by satisfying key stakeholders. If one looks at ARCO's pursuit of reformulated gasoline, or Bristol Myers Squibb's development of its Herbal Essences line, or even Monsanto's bold attempt to move into industrial products that allow sustainable development, it is possible to see why environmental leaders must operate in an increasingly public arena. Compliance might represent an expensive "three-foot hurdle" that all must jump, but new corporate product lines that are based on environmental considerations must scale the "nine-foot hurdle" of public expectations.

> A leader reviews and assesses results primarily in three areas: key appointments and promotion, results compared to the plan, and the connections to key publics. Promotion, key appointments, and succession planning are the most crucial elements in the organization's future. These activities are a leader's true domain. The organization has a right to understand the criteria used in these decisions, and each one must be examined carefully.
>
> MAX DEPREE, *Leadership Jazz*

These three goals are never reached alone. The environmental leader figures out how to find and harness them all – not in a single person, perhaps, but in the team he or she leads. The leader serves as the example that such integration of skills and goals is possible and profitable.[3]

Personal leadership skills and strategic restraint

There are at least nine skills that enable a leader to identify the right team, and then race the hurdles in record time. What follows sums up decades

> **Box 6.1 Top personal skills for environmental leaders**
> 1. Forget about blame – find out what works.
> 2. Build a broad and deep network of personal friendships, associations and affiliations.
> 3. Cultivate risk, ambiguity and uncertainty as sources of powerful change.
> 4. Select brilliant, reliable deputies.
> 5. Check your instincts against your clients' needs.
> 6. Replicate success, using lots of small steps to clear the top.
> 7. Make the future of the organization promising to everyone in it.
> 8. Use stories and metaphors to reinforce the goals of the organization and a sense of belonging.
> 9. Acknowledge the importance of everyone's role.
>
> (B.W. PIASECKI for AHCGroup, 1997)

of research into political and corporate leadership, with a crisp bias for attributes that should help those managing resources with sustainability as a strategic goal (Box 6.1).

The first is the ability to find what works, not just what is right. Leaders seem to sidestep blame, stepping back to find what works, or better, to reveal what works best. Perhaps Bill Bradley's idea about the "unlocking" of potential starts here, since environmental leaders focus themselves and others on answers, not opinions and positions.

The second important skill is recognizing the power of affiliation. National Public Radio commentator Joel Makower once stated whimsically at a conference: "Remember, there is only one letter difference between networking and not working." That letter spells the difference between creative sustained growth and institutional stagnation. In other words, there is little lasting value in a professional life without networking or affiliation-building, and environmental leaders seem to grasp this intuitively. For instance, an effective environmental leader may spend a significant chunk of the week rediscovering the values latent in his or her network of personal affiliations, from old employers to new hires, from contact with regulators who once gave the firm a slap on the wrist to new college kids looking for a break. Leadership is about viewing environmental challenges as a route with a few good options, not a downward spiral toward dead ends.

Affiliation suggest a new turn: the locked door turns out to have a key controlled by your old colleague. Affiliation is quite different from conventional sources of corporate advantage, which stem from position-based or command-based task definition. The environmental arena can best be thought of as a spider web of interactions that link people who want to join together. Note that it is most often commitment-based, not command-based, and that leaders use person-to-person influence, not just position-to-position influence. This power of affiliation is implicitly echoed in Carla Berkedal's contribution about spirituality and ethics, as well.

A tolerance for productive ambiguity is the third skill. Suppose you are working in an environmental consulting firm and that the overwhelming majority of your company's clients are petrochemical based companies. Then your leaders, sensing but not quite knowing that changes are in the wind, ask you to look into attracting clients from cosmetics and other low-volume, low risk specialty manufacturing businesses. In 18 short months, you acquire a whole new list of clients who make adhesives and other products with issues involving volatile organic compounds (VOCs).

The Clean Air Act Amendments, it turns out, had rather suddenly prompted these new clients to act, and you were ready to help them. Somehow, the leader's wider margin of imprecision allowed the team to cohere and prosper.

How did your leaders move the company toward these new clients? The answer is not direct, but as Max DePree warns in his book on leadership, "People with vision inject ambiguity and risk and uncertainty in our lives." The leader pushes his firm to find new opportunities in new places. This skill will prove critical in the choices involving sustainability, where the items of poverty, growth, and equity require productive ambiguity.

Leaders also persuade others to pursue the thrill of the chase. What enables them to survive on such high-altitude challenges? We may think we know that realistic changes are only incremental. But the leaders – such as ARCO with its reformulated gas, or AT&T with its ozone-free electronics, or Patagonia with its recycled plastic fibers – show how a tolerance for ambiguity can spawn radical, productive change.

Leaders know how to select brilliant deputies – the fourth vital skill. Frank Friedman, senior vice president of Elf AtoChem, has said: "Too many companies still employ technicians rather than managers. I would rather have one swan than two turkeys working for me" (Friedman, 1997). Selecting brilliant deputies allows you to maintain compliance as you stretch for the further goals that compound value throughout the organ-

It has been over 28 years since the birth of American environmentalism, so we all know that the search for environmental excellence is not a straight road. Back in 1990 I warned:

> The road to environmental excellence is not smoothly paved with the certainties of science, nor clearly marked by a legalistic black-and-white view of what's ahead. The desire for an easy ride in the search for answers will often be frustrated by confusion and fears. In response, the technocratic elite may propose to repave the road in order to give the issue the appearance of certainty.   (Piasecki/Asmus, 1990)

We have come a long way the last eight years, with Rio and Kyoto high on the agenda of more and more world leaders. As we enter the next century, I am happy to see that much of my warning was unnecessary. Few expect the debate over sustainable development to be paved over by scientific certainty or legalistic spin doctors, and practically no one expects the issue to take on the appearance of certainty that I had feared in 1990.

Nevertheless, we need more than ever to educate future citizens and leaders for judgments that allow sustainable options. How can this be done? In this last section, in an effort to shape a preliminary response to this looming question, I will expand my past definitions of environmental excellence by attempting to connect them with stages, or choices, we have within our own careers.[4]

While Maslow's spiral of self-actualization sums up the predicament quite nicely, the following chart (Figure 6.1) replaces the question in the commonplace terms of professional success. I do this because I suspect that it is a lot easier for a successful, "reputation-rich" professional in the prime of a career to worry about sustainable development than, say, a college freshman. One reason sustainability seminars fail in most campuses is because the subject requires a maturity of concerns more apt to be found in Elderhostel. Clearly, those that will make sustainable development real must be able to climb such a professional stairwell often, on a set of complex choices.

Leadership and its complexities in regard to sustainable development

To remind yourself of the complex and often competing goals inherent in the search for sustainable development, please see note four where we recreate the PCSD's stated ten key goals clearly. Such goals will require, and often demand, significant organizational change, along with the liberation of personal restraint among millions of professionals and

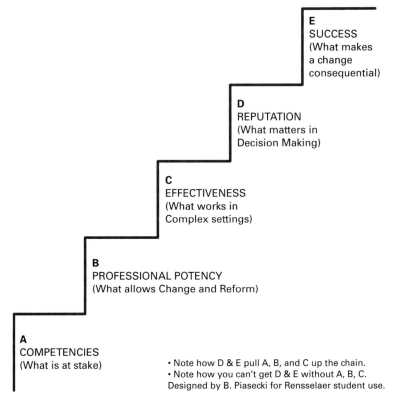

Figure 6.1 The great chain of professional expectations.

responsive consumers. John Kotter (1996) aptly notes eight common modes of failure in attempts to transform complex organizations.

In many ways, this essay's focus on leadership has been assembled to outsmart these common organizational errors. Clearly, sustainable development advocates must deflate complacency, erase probable obstacles, reassert vision – while never prematurely declaring victory. Here Cairncross's reliance on considering sustainable development as an ideal is both underdeveloped and inadequate. Take for example, Kotter's best chapter, which is on the importance of generating short-term wins. He notes, "Having a good meeting usually doesn't qualify as the kind of unambiguous win needed in this phase, nor does letting two people to stop fighting, producing a new design that the engineering manager thinks is terrific, or sending 5,000 copies of a new vision statement around the company. Any of these actions may be important, but none is an example of a short-term win." Idealists often pro-

---

Common Errors

➤ Allowing too much complacency

➤ Failing to create a sufficiently powerful guiding coalition

➤ Underestimating the power of vision

➤ *Undercommunicating the vision of a factor of 10* (or 100 or even 1000)

➤ Permitting obstacles to block the new vision

➤ *Failing to create short-term wins*

➤ *Declaring victory too soon*

➤ Neglecting to anchor changes firmly in the corporate culture

Consequences

➤ New strategies are not implemented well

➤ Acquisitions do not achieve expected synergies

➤ Re-engineering takes too long and costs too much

➤ Downsizing does not get costs under control

➤ Quality programs do not deliver hoped-for results

Figure 6.2 Common errors of organizational change. (*Source:* John Kotter's *Leading Change*)

nounce and rely upon such items that Kotter so aptly categorizes as inadequate.

"A good short-term win", Kotter boldly continues, "has at least these three characteristics: One. It's visible; large members of people can see for themselves whether the result is real or just hype. Two. It's unambiguous; there can be little argument over the call. Three. It's clearly related to the change effort." Sustainable development initiatives will need both short-term wins and victories with lasting momentum. It must be an act of compelling emotional intelligence and restraint, not just an ever-receding ideal on the horizon.

What I like about the drive for sustainability is its sophisticated demand that we rethink the traditions of personal restraint. The concept refuses to get lost in the morass of most of its critics. We know we must learn by doing. We cannot lose time in waiting for a more perfect government, a more perfect business moment for seizing the day's opportunities.

Kotter (Figure 6.2) suggests eight common errors of organizational change. His strategies to avoid them strike me as remarkably right regarding the strategies we need to adopt before we can achieve sustainability, if even only regionally. I will end with Kotter's astute warnings. It is strange, but this book by Harvard's youngest scholar ever to be tenured in the business school, gives me hope, as does art or sustained contemplation. Moreover, the more I reread the other works in this text the more I become convinced that Kotter presents a superb summary of the kinds of hurdles most organizations will need to scale before they can make sustainability real.

Acknowledgments

The author would like to acknowledge the support of Jennifer Meyerson and James P. Harrison in the research and writing of this chapter. A graduate of Rensselaer's Masters of Science in Environmental Management and Policy program, Jennifer now works for Lucent Technologies in Morristown, New Jersey. James P. Harrison is director of public strategy services and chief of staff at the AHCGroup, and a current Master's candidate in the same EMP program.

REFERENCES

Bradley, B. 1996. *Time Present; Time Past*. Knopf, New York, NY
DePree, M. 1992. *Leadership Jazz*. Jossey-Bass Books, San Francisco, CA.
Friedman, F. 1997. *A Practical Guide to Environmental Management*, 7th edn. Environmental Law Institute, Washington, DC, p. 129.
Kotter, J. 1996. *Leading Change*. Harvard Business Review, Cambridge, Massachusetts.
Phillips, D. T. 1992. *Lincoln on Leadership*. Warner Books, New York, NY.
Piasecki, B., Asmus, P. 1990. *In Search of Excellence*. Simon and Schuster, New York, NY.
Piasecki, B. 1998. *Business Strategy and Environmental Management: Leadership Skills for the Next Century*. John Wiley & Sons, New York.
President's Council on Sustainable Development. 1996. *Sustainable America: A New Consensus*. US Government Printing Office, Washington, DC.
Wildenhain, M. 1973. *The Invisible Core*. Pacific Books Publishers, Palo Alto, CA.

NOTES

1. Today's leaders in government, business and NGOs will need to change their usual roles next century. It is likely that

> more collaborative approaches to making decisions can be arduous and time-consuming (as we have learned over the past nearly there years), and all of the players must change their customary roles. For government, this means using its power to convene and facilitate, shifting gradually from prescribing behavior to

supporting responsibility by setting goals, creating incentives, monitoring
performance and providing information.

(President's Council on Sustainable Development (PCSD). 1996.
Sustainable America: A New Consensus. p. 7)

The preceding quote sums up an approach developed by the author called "moving
beyond blame", in his 1990 book.

The federal government, in particular, can help set boundaries for and facilitate
place-based policy dialogues. These are dialogues that focus on the resources and
management of conflicts of particular places or regions while giving more
opportunity, power, and responsibility to communities to address natural resource
questions that affect them directly and primarily.

For their part, businesses need to build the practice and skills of dialogue with
communities and citizens, participating in community decision making and
opening their own values, strategies, and performance to their community and the
society.

Advocates, too, must accept the burdens and constraints of rational dialogue
built on trust, and communities must create open and inclusive debates about their
future .

(PCSD, op. cit., p. 8).

2. One significant countervailing development is the noted PCSD, now led by Marty
Sprintzer. They note:

In June 1993, when President Clinton created the President's Council on Sustainable
Development, he asked us to find ways "to bring people together to meet the needs
of the present without jeopardizing the future." He gave us a task that required us
to think about the future and about the consequences of the choices this generation
makes on the lives for the future generations. It is a task that has caused each of us to
think about human needs, economic prosperity, and human interactions with
nature differently than we had before.

(Ibid., p. 2)

3. The PCSD also echoes these aspects of the new leader, when they note:

By recognizing that economic, environmental, and social goals are integrally linked
and by having policies that reflect that interrelationship, Americans can regain their
sense that they are in control of their future and that the lives of each generation
will be better than the last. Thinking narrowly about jobs, energy, transportation,
housing, or ecosystems – as if they were connected – creates new problems even as it
attempts to solve old ones. Asking the wrong questions is a sure way to get
misleading answers that results in short-term remedies for symptoms, instead of
cures for long-term basic problems.

(Ibid., p. 6)

4. Piasecki (1998) explores the links between stages of a career and the needs and
functions of next generation leaders. The PCSD captures this need for goal-based
leadership when they list the following:

The following goals express the shared aspirations of the President's Council on
Sustainable Development. They are truly interdependent and flow from the
Council's understanding that it is essential to seek economic prosperity,
environmental protection, and social equity together. The achievement of any one

goal is not enough to ensure that future generations will have at least the same opportunities to live and prosper that this generation enjoys: all are needed.

Goal 1: Health and the Environment – Ensure that every person enjoys the benefits of clean air, clean water, and a healthy environment at home, at work, and at play.

Goal 2: Economic Prosperity – Sustain a healthy US economy that grows sufficiently to create meaningful jobs, reduce poverty, and provide the opportunity for a high quality of life for all in an increasingly competitive world.

Goal 3: Equity – Ensure that all Americans are afforded justice and have the opportunity to achieve economic, environment, and social well-being.

Goal 4: Conservation of Nature – Use, conserve, protect, and restore natural resources – land, air, water, and biodiversity – in ways that help ensure long-term social, economic, and environmental benefits for ourselves and future generations.

Goal 5: Stewardship – Create a widely held ethic of stewardship that strongly encourages individual institutions, and corporations to take full responsibility for the economic, environmental, and social consequences of their actions.

Goal 6: Sustainable Communities – Encourage people to work together to create healthy communities where natural and historic resources are preserved, jobs are available, sprawl is contained, neighborhoods are secure, education is lifelong, transportation and health care are accessible, and all citizens have opportunities to improve the quality of their lives.

Goal 7: Civic Engagement – Create full opportunity for citizens, businesses, and communities to participate in and influence the natural resource, environmental, and economic decisions that affect them.

Goal 8: Population – Move toward stabilization of U.S. population.

Goal 9: International Responsibility – Take leadership role in the development and implementation of global sustainable development policies, standards of conduct, and trade and foreign policies that further the achievement of sustainability.

Goal 10: Education – Ensure that all Americans have equal access to education and lifelong learning opportunities that will prepare them for meaningful work, a high quality of life, and an understanding of the concepts involved in sustainable development.

> Accompanying the goals are indicators of progress, yardsticks to measure progress toward each goal. These indicators of progress suggest what information to look at to determine the progress that the country is making toward achieving the goals. They are not intended to be mandates for specific actions or policies, and they may change over time as the country moves toward these goals and learns more about the science and policy options underlying them.
>
> (Ibid., pp. 12–13)

For a much fuller and applied examination of how firms are beginning to pursue such complex goals, please see Piasecki's *Business Strategy and Environmental Management: Leadership Skills for the Next Century* (John Wiley & Sons: 1998). The author's co-authors for this new fifth book are Frank Mendelson, the AHCGroup's Vice President of Client Relations and the group's Director of Policy Initiatives and Press Relations, Kevin Fletcher. Mendelson is also executive director of *Corporate Environmental Strategy*, an AHCGroup quarterly journal, now available on Lexis-Nexis.

7

The role of science: guidance and service

Introduction

One of the most frustrating exercises, even in an article on "sustainability," is to try too hard to define this beast. The Brundtland Commission (WCED 1987), recognizing that growth "versus" environment was a poor framing of the environment-development debate, popularized the phrase "sustainable development." Ayres' (1996) attempts to address indices of sustainability provide a perfunctory attempt to summarize some of the literature that defines sustainability:

> There has been much academic debate on the exact meaning that should be ascribed to the term "sustainability." Tietenberg suggests that sustainability means "future generations remain at least as well off as current generations" (Tietenberg 1984, p. 33). In more formal language, the above formulation implies that sustainability means "non-declining utility." Repetto states, in the same vein, that "current decisions should not impair the prospects for maintaining or improving future living standards" (Repetto 1985, p. 16). The World Commission on Environment and Development suggests, in the same vein, that sustainable development "meets the needs of the present without compromising the ability of future generations to meet their own needs" (WCED 1987). Pearce *et al.* (1989) and Hoagland and Stavins (1992) have collected a number of other definitions in the literature.

Somewhat cynically, but probably with merit, Ayres concludes:

> Apart from being unsatisfactory from the standpoint of reflecting non-economic elements of sustainability, these definitions share another common feature: they are unquantifiable and, absent quantitative measures, unverifiable. Some are sufficiently vague as to permit contradictory conclusions as to whether or not the conditions for sustainability are being met in any particular case. This permits –

perhaps, indeed, encourages – perverse interpretations of
sustainability to justify continued business-as-usual by business,
governments and even the World Bank.

As a case in point, Daily (1997) presents many articles that attempt to
describe and cost out so-called "ecological services," such as waste recycling, climate stability, genetic resource diversity, or flood control. While
the absolute value of nature's services are arguably in excess of the foreseeable human economy (e.g., Costanza *et al.* 1996), the relevant questions are
(1) by *how much* are human activities disrupting environmental services
and, (2) in the cost-benefit paradigm, what are the marginal costs (in monetary, aesthetic or in equity terms) of environmental damages related to
the costs of mitigation policies. For example, Alexander *et al.* (1997),
noting the extraordinary degree of uncertainty in estimates of climatic
changes, damages and abatement policies, cite decision analytic surveys
in which a variety of experts are asked to provide their subjective probabilities that particular climate change scenarios would cause impacts to
the world economy of 2100 AD.

Subjective experts

It is worthwhile in the context of exploring what science can and can't
contribute to the sustainability debate to examine some of the results
from one such prominent survey exercise, Nordhaus (1994). William
Nordhaus, an economist from Yale University, has made heroic steps to
put the climatic change policy debate into an optimizing framework. He
has long acknowledged that an efficient economy must internalize externalities (in other words, find the full social costs of our activities, not just
the direct private cost reflected in conventional "free market" prices). He
tried to quantify this external damage from climate change and then tried
to balance it against the costs to the global economy of policies designed
to reduce CO_2 emissions. His optimized solution was a carbon tax,
designed to internalize the externality of damage to the climate by
increasing the price of fuels in proportion to how much carbon they emit,
thereby providing an incentive for society to use less of these fuels.

Nordhaus (1992) imposed carbon tax scenarios ranging from a few
dollars per ton to hundreds of dollars per ton – the latter of which would
effectively limit coal use in the world economy. He showed, in the context
of his model and its assumptions, that these carbon emission fees would
cost the world economy anywhere from less than 1 percent annual loss in
Gross National Product to a several percent loss by the year 2100. The effi-

cient, optimized solution from classical economic cost-benefit analysis is that carbon taxes should be levied sufficiently to reduce the GNP as much as it is worth to avert climate change (e.g. the damage to GNP from climate change). He *assumed* that the impacts of climate change were equivalent to a loss of about 1 percent of GNP. This led to an "optimized" initial carbon tax of about five dollars or so per ton of carbon dioxide emitted. In the context of his modeling exercise, this would avert only a few tenths of a degree of global warming to the year 2100, a very small fraction of the 4 °C warming his coupled economy-climate model projected.

How did Nordhaus arrive at climate damage being about 1 percent of GNP? He first assumed that agriculture was the most vulnerable economic market sector to climate change. For decades agronomists had calculated potential changes to crop yields from various climate change scenarios, suggesting some regions now too hot would sustain heavy losses from warming whereas others, now too cold, could gain. Noting that the U.S.A. lost about one third of its agricultural economy in the heat waves of 1988, and that agriculture then represented about 3 percent of the US GNP, Nordhaus felt the typically projected climatic changes might thus cost the world economy something like 1 percent annually in the twenty-first century. He added slight additional damages for "non-market" amenities like nature, and assumed damages increased with the square of temperature increase. This 1 percent figure was severely criticized because it didn't explicitly deal with damages from health impacts (e.g., expanded areas of tropical diseases, heat-stress deaths, etc.), losses from coastal flooding or severe storms, security risks from "boat people" created from coastal disruptions in South Asia or losses to wildlife, fisheries, or ecosystems that would almost surely accompany temperature rises at rates of degrees per century as are typically projected. It also was criticized from the opposite point of view, because Nordhaus' estimate neglected potential increases in crop or forestry yields from the direct effects of increased CO_2 in the air on the photosynthetic response of these marketable plants or reduced cold season deaths. Nordhaus responded to his critics by conducting a survey focused on the impacts of several scenarios of climatic change on world economic product – including both standard market sector categories (e.g., forestry, agriculture, heating and cooling demands) and so-called non-market amenities such as health, biological conservation, and national security.

When Nordhaus surveyed the opinions of mainstream economists, environmental economists and natural scientists (I am respondent   10, in Nordhaus, 1994), he found that the former expressed less anxiety about the economic or environmental consequences of climate change than the latter by a factor of about 20 (see Figure 7.1). However, the bulk of even the conservative group of economists Nordhaus surveyed considered there to be at least a 10 percent probability that typically projected climate changes could still cause economic damages worth several percent of gross world product (the current US GNP is around five trillion dollars – about 20 percent of the global figure). And, some of these economists didn't include estimates for possible costs of "non-market" damages (e.g., harm to nature). One ecologist who did explicitly factor in non-market values for natural systems went so far as to assign a 10 percent chance of a 100 percent loss of GNP – the virtual end of civilization! While Nordhaus quipped that those who know most about the economy are less concerned, I countered with the obvious observation that those who know the most about nature are very concerned.

We will not easily resolve the paradigm gulf between the optimistic and pessimistic views of these specialists with different training, traditions, and world views, but the one thing that is clear from the Nordhaus studies is that the vast bulk of knowledgeable experts from a variety of fields admits to a wide range of plausible outcomes in the area of global environmental change – including both mild benefits and catastrophic losses – under their broad umbrella of possibilities. This is a condition ripe for misinterpretation by those who are unfamiliar with the wide range of probabilities most scientists attach to global change issues. The wide range of probabilities follows from recognition of the many uncertainties in data and assumptions still inherent in earth systems models, climatic impact models, economic models or their synthesis via integrated assessment models (see Schneider 1997a, b). It is necessary in a highly interdisciplinary enterprise like the integrated assessment of global change problems that a wide range of possible outcomes be included, along with a representative sample of the subjective probabilities that knowledgeable assessment groups like the IPCC believe accompany each of those possible outcomes. In essence, the "bottom line" of estimating climatic impacts is that both "the end of the world" and "it is good for business" are the two lowest probability outcomes, and that the vast bulk of knowledgeable scientists and economists consider there to be a significant chance of climatic damage to both natural and social systems. Under these conditions – and the unlikelihood that research will soon eliminate the large uncertainties that still persist – it is not surprising that most formal climatic impact assessments have called for cautious, but positive steps both to slow down the rate at which humans

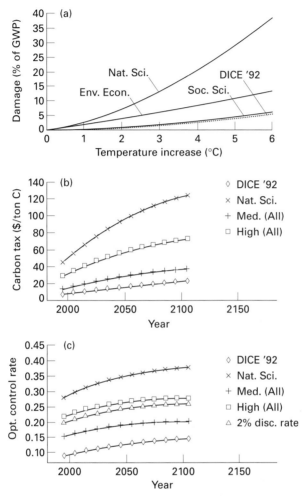

Figure 7.1 The DICE model reformulated with damage functions derived from damage estimates given by experts. Figure 7.1a shows the "disciplinary damage functions" derived from an expert survey (Nordhaus 1994), for natural scientists ("Nat. Sci."), environmental economists ("Env. Econ."), and other social scientists ("Soc. Sci."), primarily conventional economists. The original DICE damage function ("DICE '92") is also shown for comparison. In Figures 7.1b and 7.1c, optimal policy given the median damage estimates of natural scientists ($\times$) is compared with optimal policies with the median damage estimates of all the experts (+), the high 90th percentile damage estimates of all of the experts ($\square$), and with the damage estimates used in the original DICE model ($\diamond$). Figure 7.1b gives optimal carbon tax levels, for each group of damage estimates, and Figure 7.1c displays the corresponding optimal emission control rates. The increases in global average temperature by 2105 associated with these policies are 2.77 °C, 2.94 °C, 3.10 °C, and 3.20 °C, respectively. In addition, Figure 7.1c shows optimal control rates with the median damage estimates of all experts and a 2% discount rate ($\triangle$) instead of the 3% rate used in all other calculations. From Roughgarden and Schneider 1999.

modify the climatic system (e.g., Figure 7.1, b and c), and to make natural and social systems more resilient to whatever changes do eventually materialize.

Imaginable surprises

One of the reasons most of the respondents to Nordhaus' survey estimated a greater probability of very large damages than of significant benefits (so-called "right-skewness": Roughgarden and Schneider 1999) from projected climate changes is the likelihood of so-called "climatic surprises." While a "surprise" is, strictly speaking, unimagined, many unexpected possibilities can be imagined, as can conditions under which surprise becomes more likely (e.g., Schneider *et al.* 1998). In other words, the harder and faster the enormously complex earth system is forced to change, the higher the likelihood for unanticipated responses. Or, in a phrase, *the faster and harder we push on nature, the greater the chances for surprises* – some of which are likely to be nasty.

Noting this possibility, the Summary for Policy makers of IPCC Working Group I concluded with the following paragraph:

> Future unexpected, large and rapid climate system changes (as have occurred in the past) are, by their nature, difficult to predict. This implies that future climate changes may also involve "surprises." In particular these arise from the non-linear nature of the climate system. When rapidly forced, non-linear systems are especially subject to unexpected behavior. Progress can be made by investigating non-linear processes and sub-components of the climatic system. Examples of such non-linear behavior include rapid circulation changes in the North Atlantic and feedbacks associated with terrestrial ecosystem changes.

Of course, if the Earth system were somehow less "rapidly forced" by virtue of policies designed to slow down the rate at which human activities modify the land surfaces and atmospheric composition, this would lower the likelihood of non-linear surprises. Whether the risks of such surprises justify investments in abatement activities is the question that Integrated Assessment (IA) activities are designed to inform (e.g., Schneider, 1997b). The likelihood of various climatic changes, along with estimates of the probabilities of such potential changes, are the kinds of information IA modelers need from earth systems scientists in order to perform IA simulations. I turn next, therefore, to a discussion of methods to evaluate the subjective probability distributions of scientists on one important climate change issue, the climate sensitivity.

Subjective assessment of climatic sensitivity

Finally, what does define a scientific consensus? Morgan and Keith (1995) and Nordhaus (1994) are two attempts by non-climate scientists, who are interested in the policy implications of climate science, to tap the knowledgeable opinions of what they believe to be representative groups of scientists from physical, biological, and social sciences on two separate questions: first, the climate science itself and second, impact assessment and policy. Their sample surveys show that although there is a wide divergence of opinion, nearly all scientists assign some probability of negligible outcomes and some probability of very highly serious outcomes, with one or two exceptions, such as Richard Lindzen at MIT (who is scientist number 5 on Figure 7.2 taken from Morgan and Keith 1995).

In the Morgan and Keith study, each of the 16 scientists listed in Table 7.1 were put through a several hour, formal decision-analytic elicitation of their subjective probability estimates for a number of factors. Figure 7.2 shows the elicitation results for the important climate sensitivity factor. Note that 15 out of 16 scientists surveyed (including several IPCC Working Group I Lead Authors – I am scientist 9) assigned something like a 10 percent subjective likelihood of small (less than 1 °C) climatic change from doubling of CO_2. These scientists also typically assigned a 10 percent probability for extremely large climatic changes – greater than 5 °C, roughly equivalent to the temperature difference experienced between a glacial and interglacial age, but occurring some hundred times more rapidly. In addition to the lower probabilities assigned to the mild and catastrophic outcomes, the bulk of the scientists interviewed (with the one exception) assigned the bulk of their subjective cumulative probability distributions in the center of the IPCC range for climate sensitivity. What is most striking about the exception, scientist 5, is the lack of variance in his estimates – suggesting a very high confidence level in this scientist's mind that he understands how all the complex interactions within the earth system described above will work. None of the other scientists displayed that confidence, nor did the Lead Authors of IPCC. However, several scientists interviewed by Morgan and Keith expressed concern for "surprise" scenarios – for example, scientists 2 and 4 explicitly display this possibility on Figure 7.2, whereas several other scientists

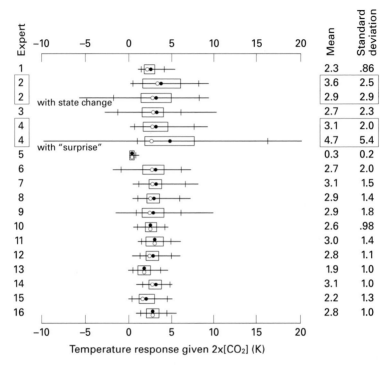

Figure 7.2 Box plots of elicited probability distributions of climate sensitivity, the change in globally averaged surface temperature for a 2×[CO2] forcing. Horizontal line denotes range from minimum to maximum assessed possible values. Vertical tick marks indicate locations of lower 5 and upper 95 percentiles. Box indicates interval spanned by 50% confidence interval. Solid dot is the mean and open dot is the median. The two columns of numbers on right side of the figure report values of mean and standard deviation of the distributions. From Morgan and Keith 1995.

Table 7.1: Experts interviewed in the study. Expert members used in reporting results are randomized. They do not correspond with either alphabetical order or the order in which the interviews were performed. From Morgan and Keith 1995.

James Anderson, Harvard University	Michael MacCracken, US Global Change Research Program
Robert Cess, State University of New York at Stoney Brook	Ronald Prinn, Massachusetts Institute of Technology
Robert Dickinson, University of Arizona	Stephen Schneider, Stanford University
Lawrence Gates, Lawrence Livermore National Laboratories	Peter Stone, Massachusetts Institute of Technology
William Holland, National Center for Atmospheric Research	Stanley Thompson, National Center for Atmospheric Research
Thomas Karl, National Climatic Data Center	Warren Washington, National Center for Atmospheric Research
Richard Lindzen, Massachusetts Institute of Technology	Tom Wigley, National Center for Atmospheric Research
Syukuro Manabe, Geophysical Fluid Dynamics Laboratory	Carl Wunsch, Massachusetts Institute of Technology

implicitly allow for both positive and negative surprises since they assigned a considerable amount of their cumulative subjective probabilities for climate sensitivity outside of the standard (i.e., IPCC 1996) 1.5 to 4.5 range. This concern for surprises is consistent with the concluding paragraph of the IPCC Working Group I Summary for Policymakers quoted above.

IPCC Lead Authors, who wrote the Working Group I Second Assessment Report, were fully aware of both the wide range of possible outcomes and the broad distributions of attendant subjective probabilities. After a number of sentences highlighting such uncertainties, the report concluded: "nevertheless, the balance of evidence suggests that there is a discernible human influence on the climate." The reasons for this now-famous subjective judgment were many, such as the kinds of factors listed above. These include a well-validated theoretical case for the greenhouse effect, validation tests of both model parameterizations and performance against present and paleoclimatic data, and the growing "fingerprint" evidence that suggests horizontal and vertical patterns of climate change predicted to occur in coupled atmosphere-ocean models has been increasingly evident in observations over the past several decades. Clearly, more research is needed, but enough is already known to warrant assessments of the possible impacts of such projected climatic changes and the relative merits of alternative actions to both mitigate emissions and/or make adaptations less costly. That is the ongoing task of integrated assessment analysts, a task that will become increasingly critical in the next century. To accomplish this task, it is important to recognize what is well established in climate theory and modeling and to separate this from aspects that are more speculative. That is precisely what IPCC (1996) has attempted to accomplish.

Complex systems, like the earth's climate and ecosystems, will be characterized by a high degree of uncertainty and technical complexity for the foreseeable future. Indeed, almost everything that's interesting and controversial, like the behavior of complex systems that involve physical, biological and social interactions, will never enjoy full understanding nor full predictive capacity, and therefore there will always contain a high degree of subjectivity. For those systems that are not objective, unlike a coin, people will have to become comfortable dealing with subjective probabilities. You can't predict skillfully the sequence of faces of a multiply flipped coin, but you can predict the odds of any sequence of faces. The probabilities of each outcome are objective and you know what they are (at least for an unloaded coin). But estimating probabilities for interesting complex systems like climate and ecosystems or socio-economic systems will involve a high degree of subjectivity – mixed in with elements of objectivity. Even for the aspects of climate for which we have lots of data and theory to make objective determinations, the way the data are applied often depends on assumptions that in turn, are subjective (e.g., Moss and Schneider 1997, 1999).

Is there a "citizen-scientist"?

What is the citizen to do in the face of this often bewildering complexity? First, citizens should learn to ask scientists the right questions. One important question that you can ask a scientist is "what can happen?" The citizen tries to get experts – whether cancer specialists, military operations officers, environmental scientists, or economists – to agree on the range of possible outcomes. Honest ones will admit that surprises are possible, both happy and unhappy, and that we have to anticipate those, too (but let's just leave our discussion here to the universe of known outcomes). A typical outcome in the case of climate change could be 3 °C warming if carbon dioxide in the atmosphere doubled.

But "what can happen" has little policy meaning by itself. What's more important is the *likelihood* of the event, so question B is "what are the odds?" Now, the probability that the Earth will be hit by an asteroid that could wipe out 50 percent of the existing species and 99 percent of living things (other than bacteria), is exceedingly low, something like 1 in 10 million per year. That's still a large number relative to DNA fingerprinting introduced into evidence in some infamous trials where 1 in 1 billion odds are asserted to be a "reasonable doubt." In other words, people have to learn to know what probability numbers mean. One in 10 million is an exceedingly small probability – that of being killed in a jet airplane crash. So, basically, if the probability is that low we usually don't do much about it. Therefore, the asteroid-Earth collision – or virtually the end of our world – remains an un-dealt-with risk. On the other hand, a probability of 1 in 100 of being killed is very high – like that often faced by inner city young men. So, the citizen interested in science has to learn quite a bit about the meanings of probability.

Now, what happens when Scientist A says the probability of some catastrophic outcome is 25 percent and Scientist B says it's 2.5 percent. Citizens, even if statistically literate, can easily get confused, especially when each scientist uses long and complex technical arguments to back up their dissimilar intuitive, subjective judgments. This not untypical situation is where *science-as-a-community* becomes important. It's very difficult for an average citizen to listen to a technical debate, and really know whose subjective opinions about the likelihood of the various assumptions that underlie disparate conclusions are more in the center of the knowledge spectrum than others.

Past episodes of basic changes in scientific thinking – "paradigm shifts" – are frequently invoked by contrarian scientists and politicians to argue that the consensus view might turn out to be false. Indeed, many famous examples in the history of science demonstrate that new discoveries, or new theories reinterpreting well-known evidence, can displace apparently well-established knowledge virtually overnight, as Einsteinian mechanics displaced Newtonian mechanics. Despite enormous progress towards stabilizing basic theories in the latter half of this century, such paradigm shifts continue. History provides no reason to suppose that many more will not occur, even if we are forced to endure more "end of" popular books sporting polemical titles like the *End of History* or *The End of Science*.

At the same time, within the history of any given discipline the number of real paradigm shifts – truly fundamental changes in concepts, methods, and conclusions, sometimes called "scientific revolutions" – has generally been quite small. Thomas Kuhn (1962), originator of the concept, observed that "normal science," which consists of extending and analyzing the dominant paradigm's observations, experiments, and models, makes up the vast bulk of scientific activity. This broader role for science in the context of social issues has been separated from the 'normal' science of Kuhn and relabeled 'post-normal' science by sociologists of scientific knowledge (Funtowicz and Ravetz 1990; see also Jasanoff and Wynne 1998). Just as in advertising, the all-too-frequent claims of "revolutionary" change are usually no more than rhetoric. Furthermore, the vast majority of paradigm challenges fail. (One recent case in point is the ill-fated idea of "cold fusion.") It is true that brilliant, revolutionary ideas have sometimes been ignored merely because they did not fit existing modes of thought, and it is virtually certain that this is happening, somewhere, now and will happen again in the future (e.g. Schneider *et al.* 1998). Yet it is also true that most paradigm-challenging ideas fail because they are – to put it bluntly – simply wrong. Just as with war, history tends to lionize successful scientific revolutionaries while fast forgetting the far larger number of failures.

Let me rephrase this in two points. First, being in the center of the current knowledge spectrum doesn't mean that you're right! Ptolemic supporters with long beards and flowing robes, holding high positions in church and state, held sway for many centuries with the wrong theory. However, most problems are not Copernican. For most problems, the scientific mainstream is not far away from truth. And even for those few problems that turn out to be Copernican, for every real Copernicus there are probably a thousand pretenders. Only rarely does someone with great convictions and correct insights stand up to the scientific establishment and overthrow the dominant paradigm. In my experience, most self-proclaimed paradigm-busters are not the genuine article – which does not mean they should be ignored, but welcomed only skeptically.

So, this brings us to the prime problem: How does the citizen figure out where the mainstream is and whose subjective probabilities to trust? The "citizen-scientist" becomes an oxymoron when facing this dilemma unless he or she is willing to put in a lot of effort: going to the library, reading National Research Council studies, Intergovernmental Panel on Climate Change assessments, etc. – in essence, trying to understand whether the subjective opinions of dueling experts represent marginal or mainstream views. Figuring out whether the marginal views are truly breakthrough (or in fact anything other than special interest disinformation) is a tough nut to crack. I was once accused in debate of advocating "science by consensus" rather than by experiment. After all, no Galileos would have emerged if only the consensus opinions were trusted. But I do believe in science by experiment (e.g., Schneider and Mesirow 1976, p. 10). All things in science are forever questioned and questionable. However, without inconsistency, I also believe in *science policy by consensus*. We've got to go with the best guesses of the majority, always keeping in mind the following three questions, and always challenging the dominant paradigm while acting as if it were still likely to be true.

The three questions

I already said question A is "what could happen?" and question B is "what are the odds?" Question C, then, is "how do you know?" In the process of asking "how do you know," you are asking to get deeply involved in the details of the debate. For example, if some scientist or surgeon tells you that there's "certain truth," and doesn't also point out the range of underlying assumptions and their uncertainties, then maybe that surgeon's or scientist's credibility should go down in your own mind. Cocksureness about complex issues is a telltale sign that it's time for a second opinion. But even that determination is a tough job for the citizen who is not very familiar with typical scientific debates. Sophisticated ability to discern who is more credible requires a citizen that is more than casual, but really is in that passionately interested group of science buffs that watches many Nova's and religiously reads *Scientific American* or the *Science N.Y. Times* or is willing to go to the library and ferret out credible views from marginal claims.

I include in that category of scientist-citizens good science journalists, or once in a while, senators and congresspersons – the few who really attend the bulk of their own hearings when scientists appear in detailed debates. I think the current U.S. Vice President is in that category, as is the former senator from Colorado, Tim Wirth, or the former congresswoman from Rhode Island, Claudine Schneider (no relation), or the late Congressman George Brown from Southern California, or former Prime Minister of Great Britain, Margaret Thatcher. Such people have learned enough of the process of science over the years to become like a good science reporter: they have a nose for phonies or for people who are passionate without balance. I don't mean "balance" in the traditional journalistic sense of pitting one extreme versus the middle or the opposite extreme without labeling where each claimant sits in the spectrum of knowledgeable opinions. Such inappropriate "balance" usually leaves nearly everybody not deeply familiar with the debate and the debaters confused.

Meta-institution building: a science assessment "court"?

Being able to judge a scientist's credibility is a tall order for most members of Congress – or even their technical staffers. So what I think we need, at the risk of sounding like an elitist, is a "meta-institution." We need somebody in between lay people and the expert community to help the citizen sort out conflicting claims. I'm not implying that such a meta-institution is to help them decide, for example, whether or not nuclear power is safe or whether we should have a carbon tax – those are personal value judgments based upon each person's political philosophy as to whether the risks and costs of one energy system justify turning to another, or whether investing current resources are worth the expenditure to reduce risks of performing the climate change experiment on Laboratory Earth (Schneider, 1997a), a gamble which might lead to enhanced extinction of species or increased severity of hurricanes. I believe that these are value judgments every citizen is already equipped to make – if only he or she knew what can happen and what the odds are.

To repeat, citizens need expert help with the "what can happen" and "what the odds are" parts of policy-making. That is, laypeople need guidance in figuring out where the mainstream expertise lies – that's where we may need some new institutions. We actually have had institutions with the goal of performing such assessments of complex issues. One such was recently killed by the U.S. Congress: the Office of Technology Assessment. It was a nonpartisan congressional assessment office whose job it was to provide reports that could cut through the claims and counter claims of the special interests, faxed by the thousands to the halls of power and the media – what I label the "one fax one vote syndrome." Too often, lobbying groups in the name of free speech buy bigger and bigger megaphones for non-mainstream experts whose "science" ostensibly supports their positions. If they get heard enough, these amplified experts might get equal status at the bargaining table with those espousing consensus views – i.e., repeated exposure often buys equal credibility for what really is not a very credible scientific position by simply overwhelming the communications apparatus and scientific literacy of most citizens with technical counter-examples (a good example of "post-normal" science).

Often, such disinformation specialists are counting on citizens not being able to sort out the credibility of conflicting technical claims for themselves. This is common; every citizen is used to it. Nevertheless, it is very baffling and confusing. Staffers at the Office of Technology Assessment were not fooled by this strategy, nor have been National Research Council Committees or Inter-governmental Panel of Climate Change experts. What those mainstream assessment groups do is to get together and evaluate the credibility of various conflicting claims. They do not decide, for example, whether we should impose a given carbon tax, because that is a personal value judgment on how to react to the probabilities of a set of possible consequences. The expert assessment teams decide *whether the claims made about those probabilities are credible* or at least what the subjective probability estimates might look like – as the Nordhaus (1994) experts did. I do not think that that task can easily be done by lay citizens.

That's where I separate out the role of the citizens from the experts. Each patient should decide whether to undergo an operation based on the pros and cons of surgery versus medical treatments versus dietary changes etc. But we certainly are unlikely to trust ourselves to estimate the probabilities that we have some possible condition, or the likelihood that various alternative treatments would be more or less effective or risky. Typically, the experts assess the odds, but we make the choices about how to take risks of various kinds.

If we set up a meta-institution, something in between the citizen and the scientist, it is absolutely essential that it have one characteristic – openness and transparency to all citizen groups, including special interests. For credibility, the assessors cannot meet behind closed doors where they would be in greater danger of scientific elitism or personal value biases creeping into their ostensibly balanced scientific judgments. Unfortunately, there are scientists who believe that it is inappropriate for scientists even to discuss in public issues in which there's a high degree of uncertainty. I think that's completely an elitist position, because what it says is that we scientists should be the judges of *when* we tell the public what is possible and at what odds. I think that openness is essential: reporters need to be there; special interests need to be there; ordinary citizens need to be there; as *witnesses to the assessment process itself* (Edwards and Schneider 2000) Citizens' roles are not to determine what the probabilities of various claims are because that's not their competency, but to make sure that the assessment process is open and asking the right questions.

A "FED for science"?

The most difficult aspect is how we choose who sits on this "science assessment court." This is not a court that assesses guilt and innocence or makes policy recommendations, but a court (see also A. Kantrowitz 1967) in a sense that it can evaluate the probability of claims. Again, at the risk of sounding elitist, I think the assessment team's members should primarily be chosen from the lists of the scientific societies (National Academy of Science, American Physical Society and so forth), perhaps with a few spots reserved for political appointees. The tenure of such appointees should survive the electoral terms of presidents and senators: perhaps a ten-year term. I guess I'm calling for a Federal Reserve Board-like institution, but my "FED for Science," unlike the actual FED which makes economic policy, is to be *only* an information agency. Its job is to label the exaggerators, the distorters, and the passionate who just can't (or won't) see past their own denial of special interest. Perhaps the best metaphor is Consumer's Union, a private watchdog and ratings service for products and services that assesses claims by producers or providers against objective and subjective tests.

As I suggested, citizens and interest groups should definitely be agenda setters and witnesses to the new institution, but not assessors. This is my concrete proposal for getting the public more involved in the sustainable development debate. If the public is totally confused by a baffling technical brouhaha, then instead of participating in the value-laden, policy choice process, they're more likely to abdicate to the experts. That withdrawal of citizens from the decision-making process leaves the field wide open to the special interests to compete over who can shout louder, take out bigger ads, whose fax machines have a higher baud rate, or who can finance which congressional representative's campaign most – all in order to propagate their special brand of "scientific truth." I would like to move the process of evaluating scientific credibility away from the political arena and into a meta-institution that has no responsibility for policy choice, has no decision-making authority, but can call anyone's statement, including one from the President or the Speaker of the House, "scientific nonsense" (not a rare occurrence, I'm afraid).

It is an act of courage for a scientist in the Executive branch to come out and say, "I'm sorry, the facts of my President are wrong." Such acts of courage are, almost by definition, discouraged by the hierarchical nature of the political chain of command; that's why we need an independent information agency (what I, somewhat tongue-in-cheek, labeled the "Truth and Consequences Branch" in my first book, Schneider and Mesirow 1976). Information cannot be under the exclusive control of people with vested interests in the answer. All interests should be witnesses to the process; they should be able to ask questions of the assessors; should be able to have their favorite "new Copernicus" able to testify to the assessors; they shouldn't be able to vote on the credibility of the conclusions.

As I stated earlier, we already have institutions that do scientific assessments. One problem for the NRC is that it is under political attack from people who claim it's "elitist" because there are no citizens in the process – other than to fund their studies. A second problem is the NRC only operates when somebody gives them enough money to do an in-depth examination of very specific questions. Moreover, the sessions are typically closed, with agendas set by the study members and the funding agency. I'm thinking of a new body that can take a letter from an ordinary citizen baffled by some conflicting Op-Ed essays, from a congressperson skeptical of a scientific witness, from an environmental group or an industrial lobby and provide a considered (say 6 weeks) reaction, not a two-year focused study. This "science assessment court" would partly rely on existing NRC studies for input, but not try to duplicate them. The assessors would basically be in the position of evaluating quickly the credibility of a whole series of claims and counter-claims about the validity of some scientific proposition or the probability that some outcome will occur. Moreover, they could commission an NRC study, or they could ask Congress to do that.

I'm not clear on how to set up this new body, nor do I envision a massive new bureaucracy, although some permanent staff are obviously needed. I'm thinking of a network of experts not all sitting in the formal institution, but who are willing on short notice, to spend several days a week for a month to write a report on the credibility of specific issues. Their meetings would be available to be witnessed either in person or perhaps by closed circuit TV. I don't claim to know what the right institution is, nor the degree to which it should be internationalized to deal with global environment-development sustainability issues. I propose the concept of a new institution because the current cacophony of claim and counter claim is disenfranchising citizens from the scientific process. If my specific proposal angers some people, then they should propose a better one. In my first book, *The Genesis Strategy*, I proposed a "Truth and Consequences Branch," a fourth branch of U.S. government, where people would be appointed for 20 years in staggered terms. But I had a much higher visibility bureaucracy in mind then, whose job was to expose the phony scientific claims of the government. Now I'm not just concerned about the scientific claims of the government. A lot of policy is made on the basis of junk science (e.g., as documented in Ehrlich and Ehrlich 1996), and some special interests are going to hate my proposal. They're likely to attack it on the grounds that it's an unnecessary federal expenditure to create a bureaucracy that will engage in inappropriate interference with their constitutional privileges to advertise their brand of "truth" as loudly and effectively as they can. After all, such defenders of infomercials will claim, "both sides" are free to promote their views. I don't propose banning their faxes or their advertisements. However, neither do I think there's anything wrong with using public financing to evaluate the scientific components of their various claims and counterclaims. The big problem for my proposed meta-institution is the standard political question: Who is it that will be doing the evaluating? Clearly, citizens need to be an integral part of that process.

Advocacy, yes; selective inattention to facts, no

I personally dislike the way the courtrooms often handle expert witnesses. It is a very bad way to ascertain the truth when each side picks elliptical experts who do not believe it's their job to make their opponent's case. And I think that's a scientifically unethical epistemology. I believe it is the job of an honest scientist to examine *all* plausible cases, and then to provide a subjective probability for each conceivable outcome that honestly reflects the range of information each expert believes to be most credible. Now, an expert could have a personal opinion of what to do with this probability assessment, which raises the question: "Is the scientist-advocate an oxymoron?" I believe that it's not an oxymoron, but requires great care. The scientist-advocate must work hard to separate out the factual from the value components of a debate. But an unconscious prejudice can be worse than a conscious one because if your prejudice is unconscious you can't even fight to fix it. At least conscious prejudice creates guilt. With unconscious prejudice or ideological zeal, advocates can be captured by their perceived brand of "truth."

To me, the best safeguard for public participation in science-based policy issues is not to leave subjective probability assessment to a few charismatic individuals, but to the larger scientific community. Some will say it's impossible for an expert to be in a public debate with policy overtones and retain his or her objectivity in the science. I think that is not necessarily true. Just because some people in the world cheat doesn't mean all do. No one is exempt from prejudices and values. The people who know it and make their biases explicit are more likely to separate them from their sober judgments about probabilities than those professing to be value neutral. If you *knowingly* distort what you believe to be the likelihood of certain outcomes for ends justify the means reasoning, you're no scientist, you're just dishonest.

However, in the real world no one – expert or lay person – gets all the time needed to explain every nuance of complex issues. We are forced to be selective or be ignored. I've called this the "double ethical bind" (e.g., Schneider 1989). To deal with the dilemma of trying to be heard but not to exaggerate, I focus on aspects of a debate that convey both the urgency and the uncertainty, typically using metaphors. For example, in the climate change debate, I could use a metaphor of a low probability of getting cancer. But I think that metaphor goes too far because if the worst happens with cancer, you die, and I don't see global warming as killing all of us or Nature – rather, I see it as a potentially serious stress that threatens selectively. A much better metaphor, therefore, would be a parasite, or a debilitating condition. But to some scientists that, too, would be an inappropriate metaphor because it's not an exact match to climate change risks. So we're always stuck on this treacherous ethical ground between finding metaphors that accurately convey the risks and uncertainties of the case, or being heard – if scientists don't find the metaphors to communicate, most citizens simply won't hear them. Instead, they'll hear the infomercials, ads, and press releases faxed to journalists everywhere by those who don't think it's their job to make their opponent's case.

When an expert communicates with metaphors and is willing to play in the sound-bite world, even though, like me, you might be uncomfortable, there is one more step you can take to be as responsible as possible. Public scientists or scientific bodies that make public statements should also produce a hierarchy of backup products ranging from Op-Ed pieces (which are often a string of written sound bites), to *Scientific American*-length popular articles that provide more moderate depth, to full-length books coming out every five years or so, basically documenting how one's views have changed as the scientific evidence changes. And even if only a decreasingly small segment of the public really wants to know what you think in detail about the whole range of questions, at least you retain an ethical stance because you have made available to anyone interested via articles and books in the scientific literature – the closest one can come to full disclosure of what's known and uncertain – which is required of honest science. But, full disclosure is simply not possible in time-constrained congressional or media debates – metaphors have to do the job, and the hierarchy of back-ups is crucial.

Rolling reassessment

What happens when the current state of science has missed something really big that's potentially dangerous, or something we currently fear proves unfounded? That's why another element to this process must always be added: what I call "rolling reassessment." It takes immediate actions to reverse long-term risks, but such actions are not without costs. Therefore, we should initiate flexible management schemes to deal with large-scale, potentially irreversible damages, and allow ourselves to revisit the issue every, say, five years. Credibility is not static – there are new outcomes to be discovered or other ones we can eventually rule out. With such new knowledge, whose credibility could be reassessed by the meta-institutions I proposed earlier, the political processes could decide we didn't move fast enough in the first place (or maybe we moved too fast) and consequently make adjustments. The trouble is that once we've set up certain kinds of large political establishments to carry out policy, people develop vested interests and become reluctant to make adjust-ments, either to the policies or the institutions. And many politicians have told me they like to "solve" problems and not be continuously forced to revisit contentious issues.

Some experts also don't like subjective assessment because they might be proved wrong. Imagine a doctor who, upon an office visit, sus-pecting that there's the possibility of a serious disease, mentions that subjective preliminary opinion to the patient, but says, of course, you must take tests. Then, when the test results come along, the physician decides to be politically consistent by not telling the patient that the tests point to a new diagnosis. That would be ultimate dishonesty. Yet, the political system seems to afford more credibility to people who predict the right answer (regardless of whether their reasoning is correct) rather than those who got the answer wrong because of factors then unknown. How unknown factors turn out is luck, not skill. The "answer" isn't really as important to a scientist as whether or not he or she gave the best judgment given what could have possibly been known at the time. Science doesn't assign credibility to people who got it right for the wrong reasons – the process is more important than the product. Science wants to know why we reach certain tentative conclusions. So should citizen-scientists.

Environmental literacy

The long-term solutions to sustainability involve more than a meta-insti-tution to evaluate credibility of conflicting claims. They involve educa-tion. They involve environmental and scientific literacy.

We rarely teach science literacy in school, even when we think we do via "science distribution requirements." Literacy is not just knowing the *content* of science, as important as that is. It isn't practical to teach detailed scientific content of a dozen relevant disciplines to all citizens – and it's not even necessary. What the citizen needs to know is the difference between a factual and a value statement; the difference between objective and subjective probabilities; the difference between a paradigm and a theory; a law and a system; the difference between a phenomenological model and a regression model (by that I mean the difference between associations of two data sets and a validated theory). Lots of people think that just because you can associate one variable in nature (or society) with another variable, and the association is predictive a few times, therefore it will always work. (That is often the right answer for the wrong reason.) But, as the cliché goes, correlation is not necessarily causation. Credible predictions come from having the processes behind causation right, not just a few correlations. When the future conditions are different than the conditions in which some correlation was first observed, a process model will likely outperform a strict empirical model. Finally, environmental literacy means knowing the political process through which decisions are made, including an ability to sort out which claims and counter-claims by the one-fax one vote folks are more credible – that's where the meta-institutions like a "science assessment court" come in.

Citizens need to have a high level of environmental and scientific literacy, but we rarely teach it. Furthermore, we rarely teach either the science or values surrounding the sustainability debate, except in a few specialty classes. I would like to see elementary schools teaching these concepts: teaching by examples and via dialogues with students, how to separate facts and values, the difference between objective and subjective probability, efficiency versus equity considerations, and conservation of nature versus development tradeoffs. It could be done. I think that environmental literacy can empower citizens to begin to pick a scientific signal out of the political noise that all too often paralyzes the policy process in dealing with sustainability choices.

REFERENCES

Alexander, S. E., Schneider, S. H. and Lagerquist, K. 1997. The interaction of climate and life. In G. Daily, ed., *Nature's Services: Societal Dependence on Natural Ecosystems*, pp. 71–92. Island Press, Washington, DC.

Ayres, R. U. 1996. Statistical measures of unsustainability. *Ecological Economics* 16:239–255.

Costanza, R., d'Arge, R., de Groot, R., Farber, S., Grasso, M., Hannon, B., Limburg, K., Naeem, S., O'Neill, R. V., Paruelo, J., Raskin, R. G., Sutton, P., van den Belt, M. 1997. The value of the world's ecosystem services and natural capital. *Nature* 387 (6630):253–261.

Daily, G. C. 1997. *Nature's Services: Societal Dependence on Natural Ecosystems*. Island Press, Washington, DC.

Edwards, P. N. and Schneider, S. H. 2000. Self-governance and peer review in science-for-policy: The case of the IPCC second assessment report. In C. Miller and P. N. Edwards, eds., *Changing the Atmosphere: Expert Knowledge and Global Environmental Governance*, in press. MIT Press, Cambridge, Massachusetts.

Ehrlich, P. R. and Ehrlich, A. H. 1996. *Betrayal of Science and Reason*. Island Press, Washington, DC.

Funtowicz, S. O. and Ravetz, J. R. 1990. *Uncertainty and Quality in Science for Policy*. Kluwer Academic Publishers, Dordrecht, The Netherlands.

Hoagland, P. and Stavins, R.N. 1992. Readings in the Field of Natural Resource and Environmental Economics (working paper). Harvard University Press, Nov.

Intergovernmental Panel on Climatic Change (IPCC). 1996. *Climate Change 1995. The Science of Climate Change: Contribution of Working Group I to the Second Assessment Report of the Intergovernmental Panel on Climate Change.* Houghton, J.T., Meira Filho, L.G., Callander, B.A., Harris, N., Kattenberg, A. and Maskell, K., eds. Cambridge University Press, Cambridge.

Jasanoff, S. and Wynne, B. 1998. Science and decision-making. In S. Rayner and E.L. Malone, eds., *Human Choice and Climate Change*, Vol. 1: *The Societal Framework*, pp. 1–87. Batelle Press, Columbus, Ohio.

Kantrowitz, A., 1967. Proposal for an Institution for Scientific Judgment. *Science* 156:763–764.

Kuhn, T.S. 1962. *The Structure of Scientific Revolutions.* University of Chicago Press, Chicago.

Morgan, M.G. and Keith, D.W. 1995. Subjective judgments by climate experts. *Environmental Science and Technology* 29:468A–476A.

Moss, R.H. and Schneider, S.H. 1997. Characterizing and communicating scientific uncertainty. Building on the IPCC second assessment. In S.J. Hassol and J. Katzenberger, eds., *Elements of Change*, pp. 90–135. AGCI, Aspen, Colorado.

Moss, R.H. and Schneider, S.H. 1999. Towards consistent assessment and reporting of uncertainties in the IPCC TAR: Initial recommendations for discussion by authors. Working draft, IPCC guidance paper distributed to lead authors.

Nordhaus, W.D. 1992. An optimal transition path for controlling greenhouse gases. *Science* 258:1315–1319.

1994. Expert opinion on climatic change. *American Scientist.* 82:45–52.

Pearce, D.W., Markandya, A. and Barbier, E. 1989. *Blueprint for a Green Economy.* Earthscan, London.

Repetto, R. 1985. Natural resource accounting in a resource-based economy: an Indonesian case study. Paper presented at: 3rd Environmental Accounting Workshop, UNEP and World Bank, Paris, October 1985.

Roughgarden, T. and Schneider, S.H. 1999. Climate change policy: Quantifying uncertainties for damage and optimal carbon taxes, *Energy Policy* 27:415–29.

Schneider, S.H. 1989. *Global Warming. Are We Entering the Greenhouse Century?* Vintage Books, New York.

1997a. *Laboratory Earth: The Planetary Gamble We Can't Afford To Lose.* Basic Books, New York.

1997b. Integrated assessment modeling of global climate change: Transparent rational tool for policy making or opaque screen hiding value-laden assumptions? *Environmental Modelling and Assessment* 2:229–249.

Schneider, S.H. and Mesirow, L.E. 1976. *The Genesis Strategy: Climate and Global Survival.* Plenum, New York.

Schneider, S.H., Turner II, B.L. and Morehouse Garriga, H. 1998. Imaginable surprise in global change science. *Journal of Risk Research* 1:165–185.

Tietenberg, T. 1984. *Environmental and Natural Resource Economics.* Scott and Company, Glenview, IL.

World Commission on Environment and Development (WCED). 1987. *Our Common Future.* Oxford University Press, New York.

8

Economic tools, international trade, and the role of business

The market sometimes fails. That is one of the calamities of capitalism. In the final years of the twentieth century, the market economy has swept the world, becoming the near-universal economic system. It has brought the benefits of choice and consumer sovereignty. It has stimulated unprecedented economic growth. But, while it has improved the living standards of millions of people around the world, the market economy offers little protection for the environment. Left to its own devices, the market delivers more pollution along with more goods and services.

To be honest, non-market economies (such as the Soviet Union) were never very good at protecting the environment either. The industrial cities left by Communism, with their heavy industries, their steel plants, and coal mines, are among the dirtiest in the world. Such protection as the environment enjoyed in the non-market nations came mainly from poverty: people were too poor to buy cars, generate garbage or build extensive new housing developments.

But market economies often deliver wealth at the expense of nature. What happens is that polluters do not carry the full cost of their actions. If a company tips waste into a river, it saves the cost of waste disposal and imposes it instead on the people who live downstream, or the municipalities that have to find other water supplies. If a coal-fired power station emits sulfurous fumes, the cost is carried by those who enjoy or make their living from forests downwind of it.

One of the biggest policy dilemmas for environmentalists is therefore how best to combine economic growth with greenery: how to rectify market failures without losing the benefits that the market also brings. The underlying assumption in this chapter is that economic growth is non-negotiable. The earth's population may well double before it stabilizes; and eventual stability is, of course, an optimistic assumption. Simply to feed, clothe, and shelter all that multitude of extra people will require continued economic growth.

In rich countries, too, pockets of poverty and deprivation remain. It would be wonderful to persuade people that the cure for poverty lay in greater redistribution of income and new social structures that encouraged more care for the unfortunate – and that merely acquiring more spending power would not necessarily solve social ills. But such a change of attitude will take many years. In the meanwhile, most voters will continue to demand that governments promise rising living standards.

So the task is to find ways to minimize the impact of continued growth on the environment. Environmentally friendly growth is a chimera: almost all economic activity does some environmental damage. In that sense, true sustainable development will always be an ideal rather than a reality. But environmentally friendlier growth is feasible. It is possible to satisfy more human wants from the same volume of environmental resources. Indeed, that is what has been happening throughout this century, as technologies become less material-intensive and as a greater share of people's expanding incomes goes on less environmentally damaging products such as education, health care, and entertainment. These are the sorts of trends that governments should encourage.

This chapter divides the task into three main parts. It examines first the economic instruments at government's command. Governments intervene in the economy in all sorts of ways: what matters is that they should choose tools that bring environmental benefits. At the moment, some government policies do enormous environmental harm, and also cost the taxpayer money. By abandoning them, governments could make two gains for the price of one: a cleaner environment and lower taxes. At the same time, those taxes that governments raise frequently penalize activities that governments want to encourage, such as saving and working hard. By changing the tax structure, governments could also win twice over: They could reduce the tax burden on desirable activities and increase it on undesirable ones, such as polluting.

Next, the chapter looks at international trade. It raises some of the most intractable policy problems. For, if one country makes its polluters pay the true costs of their damaging activities, those companies will be at a disadvantage compared with companies from less environmentally responsible nations. The rules of international trade make this problem hard to resolve: Some environmental regulations that governments can impose on their own companies cannot be extended to the products of companies from other countries. At the same time, international trade brings opportunities. It allows the export of clean technologies, and it gives rich countries some leverage over the environmental policies of poorer ones.

Finally, the chapter looks at the role of companies. In the past, environmentalists have often assumed that companies were the enemies of the environment. Many of them were, in the sense that they saw environmental responsibility as a cost, to be avoided if at all possible. Increasingly, environmentalists have come to realize that, given the right incentives, companies can make an important and positive contribution. They are, after all, the main channel through which technological innovation reaches the market; and technology has been the most powerful force for increasing the productivity of environmental resources. And companies have come to see that there may be business opportunities in environmental responsibility.

By building environmental goals into economic policy, trade policy, and corporate incentives, the power of the market can be swung behind the environment. That is the ideal. But the experience of the past decade has been only partly encouraging. Some attempts to introduce greener economic policies have foundered; the list of companies embracing green goals has grown slowly; new economies have concentrated more on boosting domestic product than on environmental protection. All this suggests that the first step towards sustainability is political will. Sophisticated economic policies to help the environment are unlikely to get out of the university economics textbook and into the real world unless that political will exists.

Economic tools

To protect the environment, governments need to intervene. But government intervention always carries a cost and frequently has unwanted side effects. So environmental policy always needs to strive to combine the maximum environmental protection with the minimum disruption and distortion to the economy.

If that is to happen, two pre-conditions are important. First, governments must try to compare the costs imposed by environmental measures with their benefits. That is not easy: measuring environmental benefits is difficult and controversial, and public opinion may worry more about imagined dangers than real ones. Secondly, governments must look for environmental measures that try to harness economic forces. In the late 1980s and the 1990s, this realization has brought a new emphasis on economic instruments for environmental policy. However, such measures have been far more discussed than implemented.

Indeed, economic tools such as green taxes are, for the moment, much less important instruments of environmental protection than either the law or regulation. So environmentalists need to take an eclectic approach. They need to look for ways to make existing environmental policies work more cost-effectively, as well as trying to make more use of economic instruments.

Making regulation work better

An important question to ask is whether regulation can be made to work better than it often does at present. Governments mainly protect the environment by telling companies and individuals what to do. Often, they set standards. Sometimes, they lay down exactly how those standards should be met: how a chemical should be tested, for instance, or what sort of equipment a company should introduce.

Economists, especially in the United States, have repeatedly pointed to the inefficiencies of regulations, and especially of the sort of detailed and rigid "command-and-control" that has typically been applied there. They point to a number of drawbacks. For example:

· When governments lay down technologies to achieve certain pollution standards or other environmental goals, they are unlikely to be as well informed as companies – and the technical standards they set will not encourage or take account of technological advance. Better, for instance, to set a standard for automobile emissions than to mandate electric vehicles.
· Regulation tends to set a floor as well as a ceiling. Companies will have little incentive to do even better than a regulation requires – unless they expect the regulation to grow tighter.
· Regulation works best when applied to a few large polluters. It has proved easier for governments to persuade the oil companies to clean up their activities than the building industry.

So why do governments go on using regulations? The answers include the fact that they are widely seen as fair; that companies know what is expected of them; and that regulation often conceals the true costs of pollution-control measures, making them appear less expensive to society than they really are. Environmentalists like such obfuscation, even though the result may be to skew environmental priorities in favor of measures that can be imposed on companies, and away from consumers.

Given this lobby, it is worth trying to make regulation work better. That can be done in several ways:

· Set the goals of regulation carefully, paying attention to costs and benefits. One famous study by America's Office of Management and Budget analyzed a number of regulations set by the Environmental Protection Agency in terms of the implicit value they put on each life saved. It found that the answer ranged from $200 000 a life saved (in the case of a drinking-water standard) to $5.7 trillion (not much less than America's GNP) for a rule on wood preservatives.
· Where possible, set standards in terms of the receiving environment (how clean should this river be, how pure should this air be?) rather than in terms of the amount of dirt discharged by a particular plant. And vary the standard according to location.
· Where possible, take account of what happens to pollutants when they are removed from the environment. It makes no sense to dump dirt scoured from the air into the water, or to incinerate dirt dug from the soil, if that merely distributes pollution through the air. Pollution rules need to take an integrated approach, asking where dirt can be disposed of least harmfully.
· Regulations, once passed, need to be enforced. That may sound self-evident. But politicians often find it easier to pass regulations than to put them into practice. Eastern European countries, in Communist days, had admirably strict environmental standards – but rarely enforced them. It may often be better to have a few environmental rules, strictly enforced, than lots of them, widely ignored.

Because so much environmental policy takes the form of regulation, it is far more important to make regulation work efficiently than to introduce new economic tools. It is almost equally important for governments to review their subsidies and to price properly the resources that they own. Typically, governments do far more environmental damage by spending money than by not spending it.

Ending subsidies

Often, the most powerful tools of environmental policy are government subsidies – but they are harmful tools, encouraging the over-exploitation of natural resources. Stephan Schmidheiny, the leader of the green business movement, has argued that the main task of governments is to demolish "the wall of subsidies and tax regimes that promote the wasteful use of resources" (Boulton 1997b).

Take three examples:

Fishing

Governments hardly ever treat the fish that swim in their coastal waters as a communal resource, charging fishermen for access to them as a landowner might charge a hunter who wanted to shoot game on his land. (A rare exception is Britain's Falklands Islands.) Instead, governments usually pour money into fishing, subsidizing fishing boats and equipment. All this spending does not create more fish, but merely ensures that existing fish are caught more quickly. Some fishing experts think that the world's fishing fleets could be cut to one-fifth their present size, and still catch as many fish as today. But as long as governments back fishermen with taxpayers' money, that will not happen.

Farming

Almost every rich country (virtually the only exception is New Zealand) subsidizes farmers in one way or another. Subsidies usually take the form of price supports. The effects include encouraging farmers to grow the crops that are best supported, regardless of whether or not they suit the land; encouraging intensive agriculture, with its accompanying reliance on artificial pesticides and fertilizers, as the best way to make money from subsidies; and fostering monoculture, as price supports mean that farmers have less need to worry about the risks that were once entailed in concentrating on a single intensive crop. Farm subsidies in rich countries allow them to undercut unsubsidized farmers in poor countries. So they spread misery in the poor world and soil erosion and habitat loss in the rich world, all at the taxpayer's expense.

Water

Most irrigation programs around the world are underpinned by government money, which pays for dams and the infrastructure of irrigation. Often, the subsidized water is used to grow subsidized crops, whose market value may actually be less than the true value of the water itself. Subsidies encourage profligate uses of water in countries that need to conserve it. Irrigation does environmental damage too: The water flushes mineral salts to the surface, where they dry out and can eventually destroy the fertility of the soil. Once again, tax revenues that could be used to do social good instead finance environmental harm.

The worst offenders here are the governments of the developed world, who account, according to estimates by the World Bank, for two-thirds of all environmentally damaging subsidies. For example, German coal subsidies account for 70 percent of the $9.5 billion given to energy producers each year by governments of the Organization for Economic Co-operation and Development. The elimination of OECD subsidies to coal producers would be enough to reduce the global output of carbon dioxide, one of the main greenhouse gases, by 1.5 percent (Boulton 1997b).

So the eradication of subsidies that encourage the over-use of natural resources is the simplest economic tool of environmental management. In addition, economies always work more efficiently if prices are not distorted by subsidy. So these are win-win-win strategies: the environment, the economy, and the taxpayer all stand to gain.

Defining property rights

Although, at first glance, property rights may not look like an economic tool, they can be a powerful weapon of environmental protection. There are two reasons.

An asset that is privately owned is usually likely to be more sustainably managed than one whose ownership is communal or unclear. Joan Robinson, a distinguished British economist, used to pose the question: "Why is there litter in the public park but no litter in my back yard?" Resources owned in common (such as forests and fisheries) are more likely to be over-used than resources whose ownership is secure and protected by law. Indeed, one of the most influential analyses of environmental damage draws on the work of Ronald Coase, a Nobel-prize-winning American economist who drew attention to the way that the owners of private property rights had a built-in incentive to protect them.

In addition, private property rights, once awarded, can be enforced by the courts. Indeed, that was the earliest way that society imposed on polluters the costs of their pollution. Many years before governments started intervening, the victims of pollution acquired legal remedies from the courts. One example is the ancient right enjoyed in England by the owner of a stretch of river bank to receive water of a certain accustomed quantity and quality and to own the fishing rights in the adjacent stretch of water. This right has been used repeatedly by fishing clubs in England to win financial damages from upstream polluters.

Such policies are being extended further. For instance, the North Atlantic Salmon Trust, an environmental body set up by fishing sportsmen in Iceland, has been buying out commercial salmon-fishing rights in Iceland, Norway, and Britain. The argument is that sports fishermen value the thrill of catching a wild salmon far more highly than commercial fishermen, who may receive only $15–20 for a fish, compared with the $1600 upwards that the sportsmen effectively pay for the ones that do not get away. By buying commercial fishing rights and extinguishing them, the Trust hopes to conserve salmon stocks for its supporters.

The concept of property rights is also influencing conservation in tropical forests and in game parks. In forests, conservation groups such as the Rainforest Foundation have worked to win government recognition of the traditional land rights of indigenous peoples. In game parks, many conservationists do not accept that local people need to be rewarded for helping to preserve wild animals. Thus Zimbabwe's Campfire project tries to ensure that local villages share the revenues and are given the meat from controlled big-game hunting, most of it by wealthy foreigners.

But obviously, there is a limit to the extent to which the environment can be converted into private property. Nobody has yet dreamt up a convincing way to give people property rights to the ozone layer, or to privatize the climate. It is precisely because so many aspects of the environment are the property of everybody – and of nobody – that they are so vulnerable to abuse.

Taxes, charges and tradable credit

In the late 1980s and early 1990s, many governments began to consider economic measures to discourage environmental damage. They did so for several reasons:

· a growing awareness of the drawbacks of regulation. As environmental regulation became tougher and more extensive, companies complained more loudly about its cost. That made governments keen to find ways to make environmental policy less disruptive and expensive.
· a realization that the most rapidly increasing aspects of pollution were caused by individuals and small firms, both of them harder to regulate than big companies.
· a search for new sources of tax revenue. Regulations yield no income for government; taxes do – and by discouraging "bads", not "goods".

Economic instruments are of two main kinds. Some, such as "green taxes" or charges for the use of natural resources, work by putting a price on pollution or on the resource that – in theory – reflects the cost that the polluter or the user of the resource imposes on society. The aim here is to get the price right. A second kind of instrument aims to hold pollution to a certain set level, by fixing the quantity that can be created (or the amount of a natural resource, such as fish, that can be extracted). The price can then fluctuate. Polluters (or fishermen) can bid for the right to emit a certain amount of pollution (or catch a certain weight of fish). These measures are known by names such as "tradable permits" or "pollution credits".

Both kinds of measure have advantages over regulation. Take green taxes: they create an incentive for a company or individual to go further than the minimum that a regulation might require. Companies that find that cleaning up is inexpensive will do so. Companies that find it expensive will pay the tax. So taxes also yield revenue. And they are fair, in a way that regulation is not: They give companies a choice over whether to pay the price of cleaning up or pay the tax.

Tradable credits have an added advantage: certainty. Because only an agreed amount of pollution is permitted, they offer the same guarantee as regulation does. But they do not necessarily yield any revenue – unless they are auctioned off to polluters, rather than simply handed out on some agreed basis, such as "grand-fathering".

In the 1990s, several governments experimented with economic measures. America's Clean Air Act of 1990 set tradable quotas for the amount of sulfur dioxide that can be emitted by power stations and other big polluters. In Britain, the government introduced a levy on the weight of garbage sent to landfills, using the revenue to reduce corporate social-security contributions. In Britain, too, John Major's Conservative government decided to increase the duty on gasoline automatically by 5 percent each year. And several Scandinavian countries introduced taxes on carbon-dioxide emissions, in the hope of stabilizing the output of global-warming gases.

The green-tax deficit

But overall, such economic instruments have so far produced much more talk than action. Why is that? Part of the problem is that environmental organizations have hesitated to jeopardize their popularity by backing fairly green taxes, such as President Clinton's attempt to raise gasoline duties in his first term. If environmentalists allow the best to be the enemy of the adequate, nothing is likely to change.

But another problem may be that the advocates of green taxes have concentrated more on the efficiencies of green taxes than on the uses that could be made of the revenues they raise. Up to now, it has been tempting to urge environmental taxes as revenue-raisers, on the grounds that finance ministries and treasury departments around the world are powerful allies. But that has a cost: it draws attention to the fact that green taxes may be green – but also take money away from people.

In the future, it might be easier to sell imports if people could see benefits, such as better public transport or cuts in income tax, that could be paid for out of the revenue. That would prevent governments from developing green taxation as a revenue source – but might make them more popular. Alternatively, green economic instruments could be applied on a revenue-neutral basis, so that they raised no extra money for government coffers, but transferred money from big polluters to small ones. An example might be a tax on gas-guzzling automobiles which was offset by a tax rebate – or even a subsidy – to the most energy-efficient vehicles. As long as revenue from taxes disappear into the bottomless pit of general government spending, new ones are unlikely to win support.

International trade

One of the biggest problems for environmentalists is that environmental standards differ from one country to another. Countries with low standards tend to be poorer, and as a result are generally low-wage producers too. Where they are successful low-wage producers, and their exports are booming, their environment is usually in a mess. Their rich-world competitors may blame higher rich-world environmental standards for a loss of markets and jobs. So questions of environmental standards become tangled up in questions of trade and competitiveness.

In addition, the environmental problems that people care more about in the rich world are often global ones – such as global warming, the protection of the ozone layer and the conservation of sea creatures such as whales and dolphins – or ones that arise in developing countries, such as the protection of tropical forests, elephants and other species. So rich-world countries often want to put pressure on poor countries to adopt the same environmental standards as they have.

In fact, it is absolutely right that environmental standards should differ from one country to another. The filthy environments of many poor countries mean that their priorities have to be basic sewage treatment, simple controls on air pollution and the provision of adequate drinking water. It is wrong to expect them to carry the cost of, say, curbing their output of global-warming gases or even protecting large mammals while thousands of children are still dying from diseases carried by dirty water. If poor countries are to meet the environmental goals that matter mainly to people in rich countries, it is reasonable for them to ask rich countries to help to pay for them. Indeed, that is the concept behind the Global Environment Facility, administered by the World Bank.

But trade between countries with different environmental standards brings opportunities as well as problems. For the technologies developed to meet rich-country standards will eventually be exported to poorer countries too. Environmentalists often talk about the transfer of technology in the abstract, as though it were something that governments could mandate. In truth, most technology is transferred either through trade or through foreign investment. When a multinational company from the rich world opens a plant in a developing country, it may not meet the standards it would face at home – but it is likely to operate to higher standards than many local plants. And the technical standards that have evolved in the rich world, for products as diverse as automobiles and electronic equipment, are likely to be adopted in poor countries too, especially if they are big exporters. Why make dirty cars for the home market and clean ones for export, especially if foreigners are allowed to sell their clean cars in your home market too?

It will not be easy to strike the right balance between the problems and opportunities that trade creates for the environment. The basic point, though, is that trade restrictions definitely make countries poorer than they would otherwise be – and do not necessarily save the environment.

The rules of trade
Many environmentalists worry that the rules of international trade undermine the ability of countries to set their own standards. The undermining comes about partly because of the way the rules are drawn, and partly because high standards affect competitiveness.

The rules seem odd to many people. Why do we need them? The answer is that, after the misery of the Depression in the first half of this century, which was caused by widespread protectionism and trade barriers, the main industrial countries formed a club in which to thrash out agreements on the rules that should govern trade. Over the years, more and more countries have joined this club, which used to be called the General Agreement on Tariffs and Trade, but is now the World Trade Organization. The rules of the WTO are agreed upon by its member countries – not imposed on them by a power-mad supranational bureaucracy. Still, the rules limit the extent to which countries can set their own environmental standards. That is something many environmentalists resent.

The main problem is a principle of non-discrimination: a rule that imported products must be treated as favorably as identical domestic ones. That allows countries to insist that imported cars have catalytic converters – as long as the same rule applies to their own manufacturers. And imported products must meet the same standards for, say, fuel efficiency as domestic goods. The important thing is that the same rules apply to everyone.

The problem arises with rules about the way a product is produced. They cannot be used to discriminate against imports if they have no effect on the product itself. Chickens may prefer laying free-range eggs to laying them in battery cages: but the eggs are indistinguishable. Fur coats made from animals caught in cruel leg-hold traps look just the same as fur coats made from animals that met a kindlier death. Tuna caught in nets that kill dolphins tastes the same as tuna caught with nets that are more sensitively designed.

This principle led to a big row in the early 1990s, as the United States tried to ban imports of tuna caught by countries (such as Mexico) which did not use special nets. It has also prevented Austria from insisting that imported timber be labeled, to say whether or not it comes from sustainably managed forests. To the naked eye, timber looks the same, whatever kind of forest it has grown in. And the non-discrimination rule also caused the European Union to hesitate to ban imports of furs from creatures caught in leghold traps (mainly from Canada).

Is the rule sensible? In other circumstances, the rule is bent: For example, countries can restrict imports of products made by prison labor, even though a doll made in a Chinese prison is the same as one made in a factory; and they can restrict some imports on health grounds, even though an apple sprayed with one pesticide may look the same as one sprayed with another. The trouble is, the environment covers so many issues. Suppose countries that had controlled the growth in their output of greenhouse gases were to impose import restrictions on goods produced elsewhere with carbon fuels. Would any trade be left?

So redrafting the rules, while desirable, is extremely difficult. For the moment, the answer is probably not for countries to try to exclude imports produced in environmentally unsustainable ways, but for companies voluntarily to opt in to certifications schemes such as the Tropical Forest/World Fisheries or some other recognized global environmental standard, such as the ISO 14000, an international quality standard of environmental management. Provided that companies can educate consumers in the importance of such standards, the market will help to prod foreign suppliers in the right direction. Clearly this is not a full solution, but at least it would be a start.

The impact on competition

Many companies complain that high environmental standards damage competitiveness. In the European Union, attempts to introduce a carbon tax were fought off partly by companies in energy-intensive industries, arguing that a tax would drive them out of business – or force them to move elsewhere. It is all very well, companies point out, to argue for polluter-pays policies; but if companies in other countries are allowed to go on dumping their pollution freely, then the principle will drive them to the wall.

In fact, it is hard to prove that high environmental standards affect competitiveness for most industries. That is a reflection, to some, of the inadequacy of environmental rules: They are simply not tough enough to impose big costs on most industries. Evidence from America's Environmental Protection Agency suggests that the cost of complying with Federal environmental regulation is on average about 2 percent for industry as a whole, although up to 10–15 percent for a few industries such as basic metal manufacturers, electric utilities, chemical manufacturers and petroleum refiners. Some of these highly regulated industries are protected by high entry costs (chemicals) and others by an element of natural monopoly (electric utilities). For most companies, though, the costs of environmental regulation are clearly well below other factors with a much greater influence on competitiveness, such as labor costs and taxation.

In addition, as work by Paul Portney of Resources for the Future has demonstrated, the difference in environmental standards among rich industrial countries is not great, especially for controls on air and water pollution (America's Superfund approach to toxic-waste disposal is unique, and uniquely burdensome). This means that two-thirds of world exports – the share accounted for by the rich industrial countries – are produced in countries with broadly similar environmental standards. Arguments about competitiveness therefore have weight only with a few industrial sectors and only with about one-third of world trade.

The worry for companies is that the cost burden imposed by environmental standards is increasing, and countries that do not care about the environment account for a rising share of world trade. There are two answers to this. One is that rich countries need to impose environmental standards carefully, being careful to avoid imposing excessive burdens on companies. It will become increasingly important to look for cost-effectiveness in environmental controls.

The other point is that, as developing countries grow richer, they will care more about environmental quality too. Here there are opportunities for rich-country companies. The industries that can improve the environment – by building waste-treatment plants, or incinerators, or upgrading garbage disposal, or offering environmental consultancy – are nearly all dominated by rich-world companies. These companies not only have a vested commercial interest in persuading developing countries to raise their standards; they have the know-how and expertise to profit when that happens. They have built their skills thanks to the higher standards set in the rich world, where 67 percent of all environmental sales are made today. But that share will decline, as developing countries raise standards. A study commissioned by the UN Development Program and published in April 1997 suggested that the global market for pollution-control equipment could more than double, to $500 billion, by 2000.

There may also be opportunities for other rich-world companies that invest in developing countries. Frequently the environmental standards they observe in their overseas plants are higher than those of local companies (even if lower than those they would be made to observe at home). Local standards may be strict on paper, but poorly and erratically enforced. Rich-world companies know a great deal about auditing and about ensuring that standards are observed throughout the company. There is, surely, scope for them to help to train the staff in the fledgling environmental agencies that are springing up in many parts of the developing world in techniques of environmental auditing. The effect might be to ensure better enforcement of environmental standards in the developing world.

Some have argued that tough environmental standards in rich countries will actually make them more competitive. That is hard to prove. High environmental standards undoubtedly impose costs on companies – which is why governments need to enforce them. But skillful companies can find ways to offset some of the burden of standards, and skillful governments can find ways to set standards that ensure the best environmental quality at the least cost.

The role of companies

The environment will not be cleaner unless companies co-operate. After all, it is companies that turn most natural resources into waste and products. Many of the things that companies do are intrinsically polluting, as are the products and the way people use them. Persuade companies to change the production process, and you have a short cut to environmental change.

In the past few years, some companies have begun to argue that responsible environmental policies are in their interest. They include big chemical companies, such as Dow, which has its own council on sustainable development; banks such as Britain's NatWest, whose board holds a special committee session on environmental issues twice a year; oil companies such as Shell, which has decided to consult environmentalists rather than fight them; and water companies such as Britain's Severn Trent, whose chief executive, Vic Cocker, believes that "If we waited for governments to start, nothing would happen." Urged on by Stephan Schmidheiny, leader of the green-business movement, a number of large firms are committed to environmental actions that take them much farther than governments have gone – or than the law requires them to go.

That is just as well. For corporate action may be a more hopeful route than trying to persuade individuals to change their behavior. Governments happily impose rules on companies that they would never dream of applying to individuals. A company may be harried for pouring chemicals into a sewer or dumping dirt with its trash, but how many governments would – or could – intervene to stop you or me doing exactly the same in our own homes?

Most governments impose environmental rules about the way goods are produced: fewer air emissions, more water treatment, better waste disposal and so on. Some set rules on the intrinsic environmental characteristics of products: the energy efficiency of washing machines, the toxicity of batteries, catalytic converters on automobiles and so on. And governments in some countries hope to persuade companies to take responsibility for their products throughout their life cycle, disposing of them when the consumer no longer wants them. All three stages create problems and opportunities for companies.

The issue for environmentalists is, first, to understand why the mood is changing among companies; and second, what can be done to encourage more change. Some environmental groups, such as Greenpeace, are starting to argue that it may be more fruitful to talk to companies than to governments. In fact, while a few high-profile companies may be willing to go a long way on their own, industry as a whole will need prodding by government as well.

Regulation and voluntarism

How far are companies likely to take voluntary action to improve the environment? Some individual companies have gone well "beyond compliance". One striking instance has been the willingness of Tesco and Sainsbury, Britain's two biggest supermarket groups, to back an international scheme for sustainable fish stocks. Another is British Petroleum, one of the world's biggest oil companies, which announced in May 1997 that it planned to work with the Environmental Defense Fund to develop international institutions to abate the world's output of greenhouse gases. BP thus became the first big oil company publicly to recognize the important link between fossil fuels and climate change.

Some countries and industries have gone even further. A survey in Britain in April 1997 found that 37 percent of a sample of 300 companies from among Britain's top 1000 claimed to have done more than the law required, or wanted to do so. That was an increase over the figure of 25 percent recorded by the survey a year earlier ("Companies compete to clean up their act" Leyla Boulton, *Financial Times*, April 15 1997).

Several government initiatives around the world have been based on a voluntary approach. For example, the National Environmental Policy Plan introduced in the Netherlands in 1989 set strict targets for reducing pollution from a few sectors of industry. Government officials worked out programs with the industries, and the results were impressive: cuts in smog, sulfur-dioxide emissions, ammonia, and packaging waste. The American government ran a similarly voluntary program in 1991 to encourage the chemical industry to reduce toxic emissions.

These instances show that voluntarism can be a powerful way to start environment action. Individual companies can seize an advantage in moral and public-relations terms. Voluntary action has other advantages. It allows companies leeway to set their own timetables for change, and to choose appropriate technological means. It allows them rightly to claim the credit for being good corporate citizens.

But, while individual companies often have admirable reasons for taking voluntary environmental action, governments should not rely on voluntarism as a substitute for legislation. For one thing, not every company will volunteer. The Dutch NEPP worked first time around because the government was dealing with sectors dominated by big companies, such as refineries. The second plan, introduced in 1993, was much less successful, because it covered sectors with many small businesses, such as retailing and building, and compliance was therefore more patchy. But big companies sometimes find that voluntary codes are not necessarily in their interest. For, if their smaller competitors ignore them, the big companies may find that they carry all the burden of industrial compliance and lose competitive edge.

Voluntary action works only if companies know that the alternative to "voluntarism" is regulation. A typical example is the agreement negotiated by Denmark's Environmental Protection Agency with Danski Industri, the main Danish industry association, in response to an international protocol demanding that Denmark reduce emissions of volatile organic compounds to 30 percent below 1988 levels by 2000. The EPA proposed a reduction of 50 percent, which it would enforce through the industrial licensing system.

Danski Industri argued that some firms would be forced out of business. The EPA therefore asked the organization to devise a program of reductions, giving no guarantees that it would accept industry's offers. In the event, Danski Industri came up with an overall reduction target of 40–45 percent, with some branches of industry promising reductions of 70 percent and others of only 20 percent. Companies that want to take part in the "voluntary" program have to join Danski Industri; those that do not will be subject to a reduction of at least 30 percent in their VOC emissions, imposed by government license (Wallace 1997).

Such a program is hardly voluntary in the normal sense of the word. But it squares up to a danger with purely voluntary programs: that some companies may call the government's bluff. In Britain, an attempt to coax the newspaper industry into a voluntary scheme to increase the amount of recycled newsprint it used has largely failed. The reason? John Major's weak Conservative government would never have risked antagonizing the powerful press over something as "trivial" as environmental legislation.

So regulation is needed. But some companies have also come to see that environmental regulation can bring benefits: not just to their reputations but to the bottom line. Once companies accept that environmental policy is going to change, they may realize that it is in their interest to push the change in a direction that is good for their business.

To be cynical, regulation is a way to exclude competitors and protect markets. Big companies are more likely to give high-profile support to environmental regulation than small ones because they know that being green costs money. They are more likely to be able to afford to invest in high standards than smaller rivals. In the waste-disposal industry, for instance, the giant Waste Management has fought to ensure the high standards for managing landfills are applied to all operators. "Cowboy" operators do not have the resources for the big investments involved.

The most striking example of this approach was the negotiation of first a reduction and then a global ban on the manufacture of ozone-damaging chlorofluorocarbons (CFCs). The Montreal Protocol and the subsequent agreements to tighten its provisions owed a great deal to the energetic support of the main companies making CFCs. What induced them to take such an apparently perverse course? The answer is that, led by DuPont, they saw that environmental campaigners would eventually force them to abandon making CFCs for the United States; also their patent protection was running out. But, to finance the expensive research needed to make less ozone-damaging alternatives, they needed the largest possible market. The best way to ensure that was to secure a treaty to phase out the use of CFCs worldwide.

Environmentalists need to join forces with companies to find other regulatory pressure points. Wherever a company has developed a new environmentally friendlier product or process, there is scope for a deal. The company has a strong interest in environmental rules designed to create a market for its wares. And, if a technology can reduce pollution, then environmentalists should surely pitch in to make sure that it is profitable too.

The pressures outside

The forces that push individual companies to go beyond compliance are often a mixture of defensiveness and moral commitment. Good environmental policies may save money at first (by cutting waste and energy use); but, as companies press on and their standards rise, they find that they are spending money. Public companies have to be able to defend their actions to their stockholders. The arguments that companies typically use include the need to reduce environmental risk, the chance to make savings, and the opportunities to exploit new markets.

A big environmental disaster can have a traumatic effect on a company. For example, Union Carbide went through immense internal upheaval in the wake of the hideous Bhopal disaster in India in 1984, and the repercussions extended to the whole chemical industry. A repeat of Bhopal may be unlikely. But the legal penalties for mistakes have risen. Fines are heavier. Company managers occasionally even go to prison for environmental misdemeanors (although not often enough, most regulators would say). In the United States, Superfund has had a powerful effect on managerial attitudes, because it is retrospective. Companies that disposed of toxic waste in ways that were perfectly legal at the time have found themselves faced with huge bills for clearing and treating it.

Companies will also be influenced by the settlement reached by American tobacco companies with the lobby of people made ill by cigarette smokers. Here is another example of companies (or rather, their customers and shareholders) paying huge penalties for damaging public health in ways that are perfectly legal. If cigarette companies can be penalized for behaving this way today, what future penalties might be imposed on chemicals or oil companies?

Sound environmental policies can often save companies money. The prototype for this strategy goes all the way back to 1975, when Joseph Ling, head of 3M's environmental department, developed a program called Pollution Prevention Pays. This was the first integrated approach within a company to designing pollution out of the manufacturing process. The plan created incentives for technical staff to modify manufacturing methods in order to prevent hazardous and toxic waste.

The program has been an extraordinary success. By reformulating products, altering processes, redesigning equipment, and recovering waste for reuse or recycling, 3M has saved $537 million. In the 15 years to 1990, more than 3000 separate initiatives allowed 3M to reduce its air pollution by 120 000 tons, its wastewater by 1 billion gallons, and its solid waste by 410 000 tons. The driving force has been the wholehearted commitment at the top of the company and the increasing incorporation of environmental issues and objectives into business planning at every level (Hawken 1993).

Many companies have copied 3M. Some claim big savings of energy and raw material costs. Most of these programs, though, have as their main goal the reduction of toxic waste. By reducing its waste output, a company not only avoids the expensive and time-consuming process of winning licenses for disposal; it also avoids the risk that it will create a Superfund liability in the future. For instance, Dow Chemical has been able to shut down several incinerators and avoid building new ones because of its success in reducing toxic waste.

The search for markets for environmental products is perhaps less likely to push firms into improving their internal environmental policies. Companies often take some time to connect their interest in the "green consumer" with the need to meet high environmental standards internally. But, once companies such as supermarkets start to promote their products as environmentally superior, they quickly find that customers want to know more about where their products came from. As a result, they begin to question their suppliers.

One company that has been through this process is B&Q, a British do-it-yourself chain, which came under attack from journalists in the early 1990s for stocking products made from tropical hardwood. The company has since sent questionnaires to its suppliers, to try to make sure it knows much more about the environmental origins of the goods on its shelves.

The pressures within

"The private sector has the ability to make change and to influence governments, if only they would pick up on this." That is the view, not of an environmentalist, but of Eileen Clausen, former US assistant secretary of state for international environmental affairs (quoted in Boulton 1997). Many people in government long to be chivvied by companies to take stronger environmental measures; just as many company managers wish that government would impose higher standards on them.

What can be done about this stand-off? First, one of the most powerful ways to change opinion in companies is measurement. Once companies start to make environmental audits, they discover possibilities and problems that they did not realize existed. Measurement seems innocuous, but it can be explosive. Companies will be more willing to measure, and to conduct environmental audits, if they are not forced to publish the results. Environmentalists should on the whole be content with that, at least as a first stage.

But measurement will be truly useful only if it continues from year to year. That is the argument for publishing a corporate environmental report at regular intervals. It allows companies to set out targets and to draw attention to the progress they are making in achieving them. It enhances credibility, especially if it is audited by an independent constancy.

Second, environmental measures will be easier to take when there is some hope of profit. The internationally agreed standard for environmental management, ISO14000, provides not just a baseline for companies that are relatively new to the business of environmental management. It also provides a legitimate way for governments to pick and choose among suppliers.

Third, companies need to realize that a good environmental reputation will help them in the most difficult market of all: the job market. As the proportion of bright young workers dwindles, recruiting the best will become a tough business, essential to a company's future. The young are the most environmentally aware of all age groups. If companies want the building blocks of the future, they will find it essential to strive for the best environmental reputation in their industry. This is perhaps the most important of all the ways in which people, and especially the young, can help to make sure that companies regard a cleaner environment as the core of their competitive edge.

REFERENCES

Boulton, L. 1997a. "Companies compete to clean up their act." *Financial Times*, 15 April.
    1997b. "Sludge and dreams at green talks." *Financial Times,* 23 June.
Hawken, P. 1993. *The Ecology of Commerce: How Business can Save the Planet.* Weidenfeld and
    Nicolson, London.
Wallace, D. 1997. *Environmental Policy and Industrial Innovation: Strategies in Europe, the U.S.
    and Japan.* Royal Institute of International Affairs, London.

9

Stakeholders and sustainable development

Sustainable development will involve a departure from the norms that have been prevalent over recorded history. This change is necessitated by the fact that human activities are much more pervasive and much more disruptive than they were historically. Herman Daly has written much about the transition from "empty world" thinking to "full world" thinking.

In many respects, an analogy to a map is appropriate. The societal maps that have been handed down from generation to generation have arbitrary boundaries and may no longer be appropriate for the types of challenges that humans are now facing. To define sustainable development is to attempt to create a new map of human interaction and direction. In the new map for sustainable development, there are certain norms that are likely to emerge and that are directly germane to social structure.

Many elements will comprise the concept of sustainable development. At the least, there will be consideration of future generations, a fusion of economic, ecologic, and community issues and the development of cooperative structures for dispute resolution and for daily living. Among these issues are implications for virtually every discipline. A key concept of sustainable development will be the integration of community and stakeholder concerns into economic and ecological concerns as shown in Figure 9.1. The goal of sustainable development is to ensure that all four elements – basic needs, ecology, ecoefficiency, and community empowerment – plus the transgenerational element are considered in the decision-making process.

This paper will explore the community and stakeholder concerns associated with sustainable development. As used in this paper, stakeholders

Community
empowerment

Meeting
basic needs

Sustainable
development
construct
across time

Ecoefficiency

Ecology

Figure 9.1 Sustainable development construct across time.

refer to the people who will be directly or indirectly impacted by a proposed facility or a proposed solution to a problem. These stakeholders
may be the citizens of a community in which a chemical plant is to be
located; they may be the indigenous people living in an area where trees
are to be harvested or oil produced; or they may be local, national or international non-governmental organizations that act to protect natural
resource and community interests.

Stakeholder concerns are commonly generated by a proposed use such
as a new manufacturing facility or some major resource project such as oil
exploration and development. Such proposals often generate concerns in
the community that is directly affected by the action. These concerns can
take many forms. There may be concerns about air quality impacts affecting public health and water quality discharges affecting the fishing upon
which the community depends. The production and disposal of hazardous waste may generate community fears. There may be concerns about

population increase associated with the new facility and resultant community growth that could affect health and safety by overtaxing the delivery of basic services. There may be concerns about the infusion of foreign cultures that could affect cultural stability in the community.

Consideration of these community concerns must be part of the formula leading to sustainable development. Virtually all parties involved with sustainable development agree that meaningful stakeholder involvement and consideration of community concerns is absolutely essential in a model of sustainable development. It is also important to note that the concepts of community and stakeholder have an element of scale to their definitions. It is easier to conceptualize and discuss stakeholders at the facility location scale. The concept of stakeholder is valid at the regional, national, and international level. However, the type of action – e.g. treaty negotiation versus project location – will change.

The integration of those who will be impacted by a proposed project into the decision-making process about that proposal is very difficult, yet very important. Historically, decisions affecting communities often had been made without the participation or concurrence of those who would be affected. A goal of sustainable development is to develop co-operative processes that will result in the meaningful participation of the affected public in decision-making. It is possible to effectively integrate stakeholder concerns with economic and ecological concerns; however, patience and effort by all involved is required.

The role of community in sustainable development thinking is certainly evident in the Rio Declaration on Environment and Development, otherwise known as the Rio Principles. The Rio Principles are the collective statement of the international community regarding sustainable development.

The Rio Declaration includes several specific principles addressing the role of social considerations and stakeholder participation in sustainable development. Principle 1 states that human beings are at the center of concerns for sustainable development and that human beings are entitled to a healthy and productive life in harmony with nature. Certainly the stability of community systems is inherent in a healthy and productive life. Principle 3 establishes that the right to development must be fulfilled so as to equitably meet the environmental and developmental needs of present and future generations, the so-called transgenerational responsibility. Principle 4 states the necessity for the integration of developmental and environmental concerns, and Principle 5 calls for the eradication of poverty, a goal addressing both economic and social concerns.

The specific issue of stakeholder involvement is addressed by Principle 10, which clearly sets out the need for participation of concerned citizens at all levels of government, including dissemination of information and effective access to judicial remedies. The specific concern for the role of women in sustainable development, set out in Principle 20, is considered a social organization proposal as well as a stakeholder provision. Similarly, Principle 22 identifies a specific role for indigenous people and their communities based upon their specialized knowledge and traditional practices. The protection of the culture of indigenous people is specifically recognized as a goal of sustainable development.

At least two major social structure themes can be distilled from the sections of the Rio Declaration quoted above – one which is more philosophical and one which is more practical. The philosophical theme is that co-operative mechanisms and equitable processes are necessary rather than structures imposed by domination. The practical theme, which reflects the philosophical, is that community concerns must be integrated into decision-making processes through a meaningful stakeholder process.

The challenge of stakeholder involvement

Stakeholder involvement is inclusive of the concepts of public participation and the provision of information to the public. However, stakeholder involvement requires more than provision of information and allowing the public the opportunity to speak about an issue. It is more than involvement in a court action. Ultimately, stakeholder involvement relates to access to and inclusion in decision-making processes, regardless of whether these decisions are made by public or private sector.

Stakeholder involvement issues most commonly arise in the context of new facility location or natural resource exploitation proposals, although stakeholder involvement ultimately is a much broader issue.

U.S. environmental law and stakeholder involvement

The experience of the U.S. legal system with stakeholder interests and involvement has been substantial and is relevant to this discussion. In some ways, these laws and interpretations offer models for meaningful stakeholder involvement, but in important respects they do not. While this discussion is specific to U.S. environmental laws, it is directly applicable beyond environmental issues. Indeed, this discussion is offered in the context of the development of a model view of stakeholder involvement in any type of decision-making process.

The NEPA model

The broadest of the natural resource statutes is the National Environmental Policy Act or NEPA. Under NEPA, an environmental impact statement is required for all major federal actions with significant environmental effects. NEPA provides the most direct example of the fusion of economic, social and environmental issues in U.S. jurisprudence. NEPA should be, and in some respects is, a model for the initiation of sustainable development processes. However, NEPA has yet to realize its potential, due in part to difficulties regarding meaningful stakeholder involvement in decision-making processes.

NEPA is an excellent stakeholder involvement tool on one level. Through the release and review of an environmental impact statement, the public is informed of proposed governmental projects, including permitting actions involving private parties. Public hearings are held in association with the release of this information, thereby allowing the public to be heard regarding certain issues. The statutory system allows access to the courts in order to redress certain types of problems, most notably a failure of the environmental impact statement to fully disclose the impacts of the proposed action to the governmental decision-maker.

NEPA fails as a stakeholder involvement tool on another level. NEPA does not mandate the protection of the natural environment or the protection of the community structure. Instead, NEPA only requires that environmental full disclosure occur prior to the decision being made. NEPA does not require that the public's input be integrated into the decision-making process in any specific manner, but only that it be heard. NEPA requires that the decision-maker be fully informed but not that the decision-making authority be shared with the impacted stakeholders.

Meaningful stakeholder involvement is attained if and when decision-making authority is shared with the affected public. Models exist for shared decision-making in NEPA processes but are seldom used unless demanded by political or community activism. These shared decision-making processes are difficult, because the political process of site and

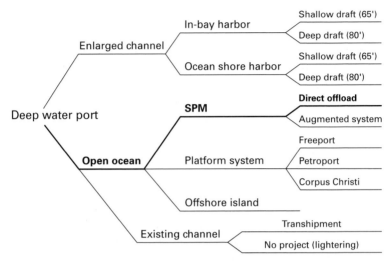

Figure 9.2 Port of Corpus Christi alternatives analysis. From Ford *et al.* (1998). © John Wiley & Sons, reprinted with permission.

project selection often preselects an alternative. Effective integration of the public with decision-making requires that this selection process be shared in an unbiased manner.

Indeed, a meaningful stakeholder involvement process will stress the identification of alternatives. The key to an effective alternatives analysis is to identify fundamentally different means of meeting project goals and objectives. Similarly, alternative locations and project sizes would be within the scope of an effective alternatives analysis. A search for alternatives is a commitment that comes with effective integration of stakeholders because stakeholders will be interested in means to alter the community impacts. A well-done alternatives analysis is an extremely powerful tool.

A well-done alternatives analysis can also expedite stakeholder involvement and the resolution of disputes over facility location and design. In Figure 9.2, the alternatives analysis diagram used in the Port of Corpus Christi deep port facility analysis is shown. This project was initially proposed as an onshore deep water port as shown in the upper portion of the diagram. This project would have cost over a billion dollars and would have involved extensive impacts to the natural environmental system as well as significant opposition that might have divided the community.

As a result of public meetings indicating a strong preference of the affected local community for offshore alternatives, the decision was made by the Port of Corpus Christi to carefully consider all alternatives including the potential of an offshore port facility. This analysis led to the determination that a single point mooring (SPM) buoy offshore facility could be constructed at a lower dollar cost with less environmental impact than was the case with an onshore port. Ultimately, the Port decided that this offshore alternative preferred by the community also was their preferred concept.

NEPA does not require such a substantive result nor does it require that the environmentally superior or community preferable alternative be selected; it only requires full disclosure of the impacts of alternatives in order that the decision-maker may consider them. This port facility example shows that the process can include stakeholders in a meaningful manner and realize its full potential.

The Clean Water Act model

The Clean Water Act model of stakeholder involvement is different from that found under NEPA. One major difference is that the Clean Water Act is substantive as well as procedural, raising new issues and challenges to stakeholder involvement. New and existing facilities that discharge pollutants into the waters of the United States must have National Pollutant Discharge Elimination System (NPDES) permits from the U.S. Environmental Protection Agency (EPA). Substantive provisions such as meeting national discharge standards and state-approved water quality standards must be followed in permit issuance. The Clean Water Act requires that a public notice be issued and that the public have an opportunity to comment upon a proposed permit. In some cases, a public hearing may be conducted in the community where the discharge is proposed.

While many stakeholder involvement goals are met under the Clean Water Act, many others are not. Meaningful involvement in the decision-making process of the agency is not guaranteed under the Clean Water Act. The presumption under the Act is that if the permit applicant meets the technical requirements for water quality protection, the permit will be issued regardless of unanswered community concerns. As an example, there are certain types of issues that are not considered at all under the Clean Water Act and its implementing regulations, including the build-up of toxic contaminants in the sediments and certain carcinogenic, teratogenic, mutagenic, bioaccumulation, and synergistic impacts of various toxic and non-conventional pollutants. If the regulations do not specifically contain requirements applicable to certain specific pollutants, then the decision to issue the permit often can be made without consideration of the public's concerns about these issues. That is the downside of substantive rules from a stakeholder perspective.

Shared decision-making

Effective methods for involving stakeholders in the decision-making process are missing from U.S. environmental law as well as from society generally. This is not to say that opportunities to participate are lacking but rather that power is not shared with the community. Sharing of power over the permitting decision and/or over the facility is the central issue of stakeholder involvement. Again, the concept of shared decision-making is applicable to problems and issues beyond a local community or facility; the community scale simply provides an excellent analytical base.

There have been many arguments advanced against more meaningful involvement by the public. One line of argument is that the applicable environmental laws currently protect the public and no further action is necessary. A second line of argument is that it is not possible to have meaningful discussions about facility location or operation with the public because the public is emotional and not oriented toward "sound science". To the extent that a dialogue with the public occurs, those holding the dominant position in the relationship often try to control that dialogue by limiting the scope and content of the dialogue.

Fortunately, there are examples of the successful integration of stakeholders into facility performance issues. In the case of Formosa Plastics in Point Comfort, Texas, a meaningful stakeholder involvement process has been successfully implemented. This process led to integration of stakeholders into corporate decision-making that affected the community. This process incorporated sound science and environmental engineering into the stakeholder involvement process. In many respects, this Formosa Plastics process is a model of how a successful stakeholder partnership can be developed.

Formosa Plastics: a case study in shared decision-making

Three innovative agreements for stakeholder involvement have been executed by Formosa's Point Comfort complex: (1) the Blackburn–

Formosa Agreement that established an arbitration process associated with the operation of the Point Comfort chemical process, (2) the Wilson-Formosa Agreement that established a mediation process associated with the zero discharge of wastewater into Lavaca Bay, and (3) the sustainable development agreement signed by both Blackburn and Wilson that initiates a comprehensive sustainability analysis of the Point Comfort facility.

Background to the agreements

In the early 1980s, Formosa Plastics constructed vinyl chloride monomer (VCM) and polyvinyl chloride (PVC) plants on the Texas coast at Point Comfort, Texas. During the latter part of the 1980s, Formosa announced plans for a substantial expansion of this plastics complex to make it fully integrated by adding seven major chemical plants and associated storage. This expansion represented an investment of over two billion dollars and thousands of construction and permanent jobs.

This proposed expansion was controversial from its inception. The Governor of Texas and other elected politicians were uniformly in favor of this proposal. However, many of the persons directly affected by the proposed expansion were concerned about the impact of the expansion on air and water quality. The town of Point Comfort was within a few thousand feet of the proposed expansion site and the proposed wastewater discharge was into Lavaca Bay, an important estuary that was already designated a superfund site due to the discharge of mercury into the bay by the neighboring Alcoa aluminum plant. Many of the pollutants to be discharged into the air and water were hazardous at certain concentrations, and provided the basis for disquiet at the community level that was exacerbated by Formosa's environmental compliance record.

The local community was unorganized until a local commercial fisherwoman named Diane Wilson formed a citizens' organization called the Calhoun County Research Watch along with recreational fisherman and environmental lawyer Jim Blackburn, the author of this article. Wilson, Blackburn and the Calhoun County Resource Watch, opposed air pollution construction permits for the initial stages of the proposed expansion as well as a second tier of additional air pollution permits. Wilson, Blackburn and the group opposed the wastewater discharge permits at the state and federal level and filed suit in federal court against the U.S. EPA and Formosa Plastics.

During the course of these legal battles, Formosa Plastics' environ-mental performance became a major issue. Formosa had numerous viola-tions including citations for (1) air violations involving releases of vinyl chloride, a hazardous air pollutant, (2) an emergency air pollution release of hydrochloric acid that shut down a road and scared the local commu-nity, (3) violations of federal and state hazardous waste provisions and (4) violations of state wastewater discharge permits. As a result of continued legal opposition and two hunger strikes by Ms. Wilson, Formosa's poor compliance record became well known and well publicized throughout Texas and the United States.

Despite Formosa's environmental compliance record, the company obtained all necessary environmental permits over citizen opposition. In fact, construction of the facility commenced with the concurrence of state and federal officials even though the wastewater discharge permit had not been issued by either the state or the EPA. The political and legal pro-cesses simply would not stop the construction of a two billion dollar facil-ity employing several thousand workers. However, opposition remained to final permit issuance.

In 1992, Formosa approached Wilson, Blackburn, and the organiza-tion with a settlement offer. If the opposition to the expansion of the plant was dropped, then Formosa would sign an unprecedented agree-ment that would allow extensive stakeholder involvement in the opera-tion of the Formosa facility. With the concurrence of Ms. Wilson, Jim Blackburn signed an initial agreement with Formosa Plastics in 1992 fol-lowed in 1994 by Ms. Wilson's agreement with Formosa. It is these two agreements and the 1997 agreement on sustainable development that implement shared decision-making in the operation of the Formosa Plas-tics' Point Comfort facility.

The Blackburn–Formosa agreement

The Blackburn–Formosa agreement created a process for stakeholder involvement in the operation of the Formosa Plant. At the time this agree-ment was signed, the most important outstanding issue regarding the operation of the Point Comfort facility was the environmental compli-ance record of the facility and resultant air pollution, water pollution, and hazardous waste issues and became the focus of the Black-burn–Formosa agreement.

The solution established in the Blackburn–Formosa agreement was to

use of a new biological treatment facility and a recycle of steam condensate.

Additionally, the studies under this agreement identified a conceptual alternative that could lead to zero discharge of wastewater. As shown in Figure 9.10, this alternative includes the existing treatment processes, utilizes the recycle/reuse alternatives identified above (the adopted processes), and proposes that the remainder of the effluent be subjected to reverse osmosis in order to allow the reuse of the remainder of the wastewater.

In addition to the 32 percent reduction in wastewater generated, the Wilson-Formosa agreement has resulted in water savings of 2.6 million gallons per day. In a water-short area such as Texas, these savings are substantial. A small portion of the water saved by this effort will be dedicated to the creation and operation of a several hundred acre wildlife habitat and wetland area which will also be used for experimental purposes with regard to wastewater treatment.

Formosa Plastics and Ms Wilson have not reached closure on the utilization of reverse osmosis or other ultimate processes for zero wastewater discharge. Rather than forcing the resolution of all outstanding issues at this time, Ms Wilson and Formosa agreed to continue the zero discharge analysis under a sustainable development agreement with the goal of zero generation of pollution from the facility. This work will include a special analysis of the brine stream at the facility, which provides the highest level of difficulty to treat.

By utilizing co-operative processes and by attempting to integrate the concerns of a local fisherwoman into plant design, Formosa has made a substantial movement toward zero discharge of wastewater. Here, the concerns of the fishing community about the impact of the plant are being addressed in a meaningful manner that would not otherwise be required by federal or state law.

The Sustainable development agreement

In 1997, Jim Blackburn and Diane Wilson signed a new agreement with Formosa Plastics' Point Comfort facility called the "Sustainable Development Agreement". Under this agreement, Ms Wilson and Mr Blackburn will represent the public interest and Formosa will represent itself. Again, a third party neutral will be selected to the Technical Review Commission, which will be responsible for commissioning studies and

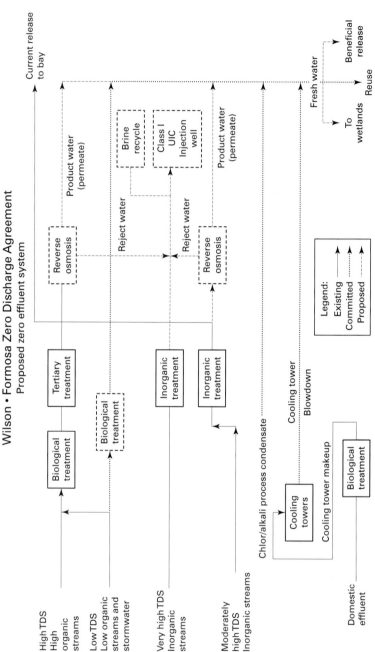

Wilson • Formosa Zero Discharge Agreement
Proposed zero effluent system

Figure 9.10 Wilson–Formosa zero discharge agreement – proposed zero effluent system. From Ford *et al.* (1998). © John Wiley & Sons, reprinted with permission.

analyses of sustainable development concepts for the Formosa Plastics Point Comfort facility.

There are many issues to be covered under this agreement. The goal of zero emissions will be pursued as will a life-cycle analysis of the various production processes utilized in the Point Comfort facility. In addition to continued compliance and pollution prevention, issues such as the flow of materials and energy through the plant will be evaluated and alternatives to reduce these throughputs will be pursued. In many respects, the concept of development identified in the work of Herman Daly will be explored in the Formosa sustainable development agreement. Of particular concern will be the flow of materials and energy through the individual plants and the complex as well as the more general flow of plastics into society. Ultimately, the issue of product and/or raw material substitution must be broached.

As shown in Figure 9.11, sustainable development thinking may be analogized to peeling an onion. An overall philosophy must exist at the outer layer of the onion. In the case of Formosa's Point Comfort facility, meaningful stakeholder involvement is a key aspect of that philosophy. The various layers include the issues that were the subject of the various agreements, including environmental compliance and pollution prevention (e.g., zero wastewater discharge). The new areas to be studied under the sustainable development agreement include material and energy efficiencies and product and/or raw material substitution where appropriate. Ultimately, the question of what is "enough" must be broached. Although we are not clear on the final interpretation of enough, it is intended to convey a limitation, a saturation point for production and consumption.

Formosa case study summary

These agreements by Formosa Plastics represent a model for the meaningful involvement of stakeholders in corporate decision-making. These agreements represent a true co-operative effort by the company and those affected by the company's action to act together to address mutual problems.

These agreements worked because Formosa Plastics empowered the stakeholders. Rather than taking a legalistic position that was allowable under federal and state laws, Formosa chose to share power with those affected by their actions. A process of accountability and a process for developing a sound scientific basis for decision-making were created by

Sustainable Development Agreement
Sustainable development framework – Facility viewpoint ecoefficiency

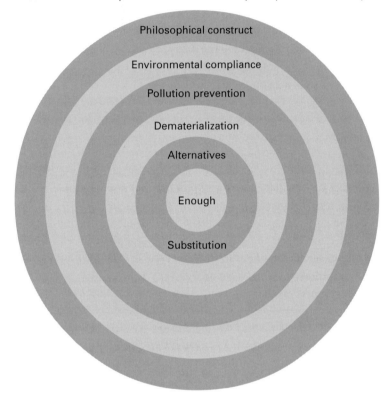

Figure 9.11 Sustainable development agreement.

these agreements. A basis for mutual respect and trust was established that endured.

The success of these agreements does not mean that all issues with Formosa Plastics have been addressed. Success at a major chemical complex requires continuing vigilance; the plant cannot rest on its laurels to date. Additionally, these agreements do not apply to the remainder of Formosa's facilities in the United States or to international operations. However, the success of the Point Comfort facility is contagious. Formosa has received ISO 14001 certification at all of its U.S. facilities and is considering the implementation of sustainable development concepts at many other locations. The actions at the Point Comfort facility are clearly a model for stakeholder involvement.

Procedure and shared decision-making

The procedures set forth in the Blackburn–Formosa and Wilson–Formosa agreements established processes of dispute resolution. One utilized arbitration and the other mediation. The Sustainable Development agreement will also use mediation. In all cases, these processes support and enable shared decision-making.

One of the major concerns in drafting these agreements was to ensure that the procedure led to and/or supported shared decision-making rather than supplanted it. Oftentimes, too much focus on procedure can obscure and hinder effective public involvement.

Procedure can sometimes appear to empower the public without actually sharing the power with the public. One example of such a situation would be community involvement processes where the meetings are held with citizens but the facility controls the flow of information. Here, processes exist but the end result is control of the flow of information and management of the public rather than a true sharing of decision-making or the creation of a meaningful opportunity for participation.

This is not to say that procedural structures and reforms are not useful. When participatory concepts are being initiated, it is extremely important to create procedures to expedite implementation. Access to information and decision-makers are important first steps. However, the important point is that procedures should not be substituted for substantive involvement in decision-making. The worst stakeholder situation is when procedures that prevent substantive involvement of the stakeholders are touted as important breakthroughs. At this point, the procedures are counterproductive to meaningful involvement. Unfortunately, we have seen many situations such as this in the United States. It is the real risk of emphasizing procedures.

Substance and shared decision-making

In the Formosa case study, a co-operative process of shared decision-making addressed substantive concerns. In order to empower stakeholders, the substance of certain issues must be addressed. This is difficult and requires a commitment to innovation and perseverance.

The first step in any meaningful stakeholder involvement process is to determine the major issues of concern. Oftentimes, concerns that are not addressed by the existing legal framework are most worrisome to the affected community. On the other hand, the only basis for opposition may be legal processes associated with issues that are of less or marginal concern. Given no other opportunity for meaningful involvement, stakeholders will utilize any entry point in an attempt to influence the decision-making process. Oftentimes a targeted discussion to identify and address the key issues of concern is important.

Any stakeholder discussion must address alternatives. If a facility is proposed to be located adjacent to a community, the initial questions are always: why here? Why this type of process? Why not a less damaging alternative? To commit to meaningful stakeholder involvement is to commit to a philosophy of serious examination of various alternatives. Similarly, the decision-making process associated with those alternatives must be revealed and openly discussed.

Trust is a major issue in stakeholder discussions. Given that substantive issues must be discussed in meaningful stakeholder discussions, a key issue becomes the selection of the technical team to work on important issues. Here, it is important that the stakeholders have an opportunity to be involved in the selection of the technical team that will be depended upon for technical information. Trust will not exist at the inception of a stakeholder involvement process. To be effective, a decision-maker should assume that trust does not exist and establish processes that can promote trust over time. If one party starts off by requiring the other party to trust them absolutely, failure is virtually guaranteed.

Oftentimes members of the technical and scientific community are most experienced at working for corporations or technical managers and are poorly skilled in working with citizen groups and affected stakeholders. Similarly, corporate managers are accustomed to deal-making and decision-making whereas many community activists do not consider themselves "deal-makers". Indeed, they are responding to a perceived threat to their community and often believe that they or their families' well-being or their economic interests are at stake. It is important to keep this perspective in mind when working with stakeholders to establish cooperative systems and partnerships.

Citizens have an amazing ability to grasp technical issues, particularly if they are working with professionals who understand how to work with citizens' groups. The goal of this professional involvement should not be to dominate public concerns by overwhelming the public with technical information but to work with the stakeholders to aid them in understanding the issues and the technical concerns.

Ultimately, one of the most important roles of scientists and professionals involved in sustainable development will be to provide trustworthy, honest, and relevant opinions to stakeholders. We must develop substantive processes that the stakeholders can trust if we are to fully integrate community concerns with ecologic and economic concerns in decision-making.

Co-operation and domination

Sustainable development is a concept with many formidable challenges to its implementation. Among the more noticeable of these challenges is the effective integration of stakeholders into decision-making structures. At least in part, effective stakeholder participation involves the substitution of patterns of co-operation for patterns of domination, a most difficult task.

Patterns of domination are evident in human systems and human society. One of the most noticeable patterns of domination is the relationship of humans with the natural environmental system. Some argue that Genesis 1:28 gave humans dominion over nature, thereby creating a right to subjugate the natural environmental system and take that which is desired. More recent Biblical scholars have argued that dominion should be interpreted as a trustee relationship between humans and the Earth – the creation that is to be maintained. Just as Biblical scholars are involved in reinterpretation of key passages, so are others reconsidering the role of domination in society.

Aldo Leopold stated that an ethical relationship between humans and the natural system, which he called the land ethic, is an ecological necessity and an evolutionary possibility. Similarly, the establishment of cooperative systems, either with the natural environmental system or communities, is an evolutionary possibility. The realization of this possibility will be the bloom of sustainable development.

The Rio Principles and global co-operation

The Rio Principles represent the consensus of the nations of the world regarding sustainable development. In addition to the sections discussed earlier under social structure and stakeholder involvement, the document contains several principles that directly address the issues of global co-operation. At the least, these principles offer a vision of a future world that is quite different than the reality of the world into which I was born.

The need for global partnership to address the disparities between the developed and less developed portions of the world is set out in Principles 6 and 7. Principle 6 calls attention to the special situation of developing countries and asks that their concerns be given special priority. Principle 7 calls upon the states of the world to co-operate in the spirit of global partnership to conserve, protect and restore the health and integrity of the Earth's ecosystem. Additionally, Principle 7 asks each of the states to assume common but differentiated responsibilities, based upon the level of consumption, the financial resources available, and technology available for use to the countries.

At the least, Principles 6 and 7 call for an equitable allocation between the North and the South as a component of sustainable development. These concerns will ultimately be one of the most important challenges of sustainable development and will test whether in fact co-operative structures can be developed to respond to sustainable development challenges.

Perhaps the most visionary references to co-operative systems in the Rio Principles are those related to peace and dispute resolution. Principle 24 states that warfare is inherently destructive of sustainable development and Principle 25 states that peace, development, and environmental protection are interdependent and indivisible. These two principles are truisms on one level but on another level represent a vision of a sustainable world that is quite different from the twentieth century. Principle 26 states that dispute resolution concepts should be integrated into sustainable development and Principle 27 states that good faith co-operation and a spirit of partnership should surround the implementation of these Rio Principles. These last four of the Rio Principles establish that alternative dispute resolution structures are necessary for global stability and that co-operative systems must be developed.

The co-operative society

It is clear that the Rio Principles envision a co-operative society. Meaningful stakeholder involvement requires the implementation of co-operative structures. In many respects, the realization of sustainable development will be a struggle to set aside our past patterns and practices and to adopt new ones. Change is inherent in sustainable development. In no area will that change be more obvious than in our thinking about domination and its role in society.

Riane Eisler has written a powerful book about domination and co-operation titled *The Chalice and the Blade*. This book traces the historical and pre-historical human record with respect to patterns of domination. Ms Eisler argues that we are currently living in a "dominator system" that can be traced back several thousand years to the end of the neolithic era. Our recorded history is a chronicle of patterns of domination that have led to slavery, colonialism and epic wars, ultimately involving the entire world.

Arguably, the dominator society has reached its evolutionary end point. Humans now have atomic and chemical weaponry that can destroy human civilization, making certain types of warfare irrational. This dominator mentality is pervasive in human society and provides an underlying basis for concern about the relationship of humans with the Earth just as it provides an underlying basis for concerns about human interactions with each other. The defining challenge of sustainability will be the transformation from a domination-based society to a society patterned on co-operation and partnership.

Interestingly, the capitalist model of economics has emerged from the twentieth century as being clearly preferable to centrally planned economies. Capitalism is based upon competition, which can be a key element of domination. The difficulty of the transition into a sustainable society will be to fuse the co-operative requirements of community and ecology with the competitive requirements of economics.

The meaningful integration of stakeholders into decision-making is at the center of this transition from a dominator society to a co-operative society. To include those without power in a decision-making process from which they traditionally have been excluded is a major step toward transition to a co-operative society. Ultimately, these stakeholder processes must exist for the global community as well as for the local or regional community.

Conclusion

Sustainable development is about change. Change is implemented by a series of small steps rather than by large single steps in most situations. The attainment of a co-operative society will take time. The important point is that the change must begin.

REFERENCES

Eisler, R. 1988. *The Chalice and the Blade: Our History, Our Future*. HarperCollins, NY.
Ford, D., Blackburn, J. & Mounger, K. 1997. *Wilson–Formosa Zero Discharge Agreement*. Formosa Plastics, Point Comfort, Texas.
    1998. Case study. *Environmental Quality Management Journal*. John Wiley.

10

From idea to action: The role of policy

Is development, as we know it, unsustainable? If so, can it be made sustainable? Can regions, nations, and the world as a whole reach a durable balance between economic, social, and environmental priorities? As we prepare for the twenty-first century, what role in reaching this goal can be played by the market and what must the state do? These are central questions in the debate on sustainable development. They are addressed, directly or indirectly, in most of the preceding chapters. Different answers emerge. I will attempt to summarize them in two contrasting views, one radical, the other incremental.

The radical view argues that sustainable development must be development without quantitative growth, the opposite of development as we know it. Why is radical change needed? Because business as usual will exhaust natural resources, overload the carrying capacity of natural systems, land, air, and oceans, and eventually lead to the collapse of human societies worldwide. The superbly successful engine that drove development since the beginning of the industrial revolution more than two centuries ago now carries in it the seeds of destruction. Maximizing income, profit and consumption are becoming counter-productive. Living in a full world the same way we lived in an empty world will be suicidal.

The radical perspective believes that sustainable development can prevent this outcome. It offers a prudent alternative to self-destruction. It does so by respecting the needs of coming generations, balancing economic with environmental concerns, reducing consumption in rich countries, and alleviating poverty in poor countries. These measures, once widely implemented, will guarantee the survival of the species. Sustainable development, in the radical view, means no more growth in population, consumption of raw materials, or pollution. "Small is beautiful – Less is more." To reach this goal people must adopt radically different values, mindsets, and patterns of behavior. Profit maximization with the help of the invisible hand must be replaced by the search for improved quality of life. The goal of individual enrichment needs to make way to communal solidarity, coherence, and sharing.

This position is strongly embraced by Herman Daly. He argues that mankind must stop the dynamics of the industrial revolution and replace it with a tightly controlled no-growth development strategy. Smaller populations, less dependence on non-renewable resources, more austere living standards in the North, and reduced poverty in the South – these are his goals. The state must intervene to reduce growth and enable a smaller but sustainable human society. There is no alternative, because the laws of thermodynamics do not allow for growth to continue at present rates. Carla Berkedal espouses this position and takes it into the religious and ethical spheres. She calls for a return to early Christian values or equivalent commitments to nature and neighbor in other religious traditions. James Blackburn also sees a need for a new ethics with focus on the main goals of sustainable development: consideration of future generations; fusion of economics and ecology; and co-operative relationships among people. Robert Kates carefully explores a strange dichotomy. Defusing the population bomb is within sight. This problem has been studied with great care, and changes in behavior and policy are leading to a stable world population. On the other hand, we avoid to understand, research, and control the continuous growth in consumption. Ultimately, he argues we have to tame our insatiable appetite and move from more to enough.

A second group of authors views sustainable development as an incremental course correction. It may be difficult to accomplish but is attainable within the existing framework of economic and social values and institutions. These writers reject the no-growth ideal, either in principle or at least for another generation or so, while today's world population grows from 5.8 to about 10 billion. Because this demographic growth is already programmed, it must be accompanied by growth in food, housing, transportation, and other sectors of the economy. Otherwise mass starvation and social disintegration will result.

Malcolm Gillis and Jeffrey Vincent argue that market incentives are needed to meet the massive development needs of the future. They believe that sustainable development is compatible with the existing economic system, because current problems are largely caused by market and policy failures. They are the result of outdated values and policies, and have become counter-productive. Bad incentives, not growth per se, account for destruction of natural resources and the ecosystem. Once these mistakes are corrected, the market economy will function satisfactorily. In the incremental view, market and policy failures are the result of wrong or outdated information and decisions. The political system, in rich as well as poor countries, can overcome these deficiencies if only it musters the strength to toss out price support systems and subsidies that distort markets and have long outlived their purpose. Better information and policies will allow the market-driven economy to feed, house, and employ the larger world population without creating an ecological disaster. National self-interest, with proper regard for the long-term goals of a country, will send the right market signals.

Frances Cairncross suggests win-win measures that corporations can take to advance their business goals while at the same time adhering to the environmental principles of sustainable development. Blackburn, notwithstanding his views about the need for fundamental changes in values, believes in the success of pragmatic step-by-step changes in corporate behavior. He relates how such measures were possible at Formosa Plastics through incessant but non-violent pressure by the parties suffering from deteriorating environmental conditions. In the end, both sides were willing to negotiate mutually acceptable solutions. The victims got relief, the company found strategic advantage in the deal, and progress toward sustainable development was made.

Only time will tell who has the better crystal ball. The purpose of this chapter is to place the two views of sustainable development, radical and incremental, in the context of policies that have been taken or proposed in pursuit of sustainable development. Political science textbooks divide the process of policy development in stages. *Agenda setting*: new ideas come to the forefront, proposals are made, and public debate is engaged. *Policy formulation*: laws and regulations are passed, treaties negotiated, and new institutions and policy instruments are built. *Implementation*: resources are allocated, programs initiated and the sum of diffused actions builds momentum. *Evaluation and re-formulation*: results are measured and learning from experience helps to negotiate course corrections. In the following sections I discuss policy development for sustainable development during the initial stages of agenda setting and policy formulation. The process is still in its infancy. I then ask whether we can learn from history by looking at policy development in response to the industrial revolution. I conclude with comments on the prospects of sustainable development to become a new policy paradigm.

Agenda setting

Until 1987 few policymakers talked about sustainable development. Only managers of renewable natural resources, fisheries, forests, or grazing land, used the terms sustainability or sustainable yield. They sought to define the carrying capacity of the resource and its ability to produce a sustained yield over time. They then used this information to formulate sustainable management practices. Both ecological and economic aspects were considered, but nobody had made the attempt to apply the yardstick of sustainability to a broad range of human activities.

This is what the World Commission on Environment and Development, convened by the General Assembly of the United Nations, set out to do. The Commission's report, *Our Common Future,* introduced sustainable development as the new policy paradigm for the twenty-first century. Writing in 1987, several years before the end of the Cold War, the Commission presented both a diagnosis of what was unsustainable and a prescription for setting things right. In the Commission's view, a new set of global challenges calls for equally new policy responses. The sum of these responses makes up sustainable development. The Commission compares the challenge mankind is facing at the transition to the twenty-first century, and the solutions it proposes to the post World War II challenge and response: avoiding a nuclear holocaust between the opposing superpowers through fear of mutual destruction and global containment. Because *Our Common Future* is the common reference point in the debate on sustainable development, I review it in some detail.

At the outset, the report describes the nature and cause of the new challenges that society faces:

> Over the course of this century, the relationship between the human world and the planet that sustains it has undergone a profound change. When the century began, neither human numbers nor technology had the power to radically alter planetary systems. As the century closes, not only do vastly increased human numbers, and their actions supported by mighty technologies, have that power, but major, unintended changes are occurring in the atmosphere, in soils, in waters, among plants and animals, and in the relationships among all of these.
>
> (United Nations WCED 1987, p. 343)

Two important points are made here. First, the new challenge is not caused by some hostile force but by ourselves. It seems that we must now pay a price for the inventions and innovations that defined the industrial age, enabled unprecedented demographic and economic growth, and improved standards of living for millions of people. Second, the nature of the challenge is unprecedented. Its reach is global, the change to physical systems may become irreversible, and the impacts on human societies are unknown. It is unclear, therefore, if history will teach us much in the current situation. Nor do we know how much time we have for the lengthy process of social learning and experimentation to deal with the challenges of the twenty-first century. The stakes are high, comparable only to the threat of nuclear holocaust.

The report, *Our Common Future,* then identifies three interrelated manifestations of the twenty-first century challenges. *Survival of the human species*: Mankind, for the first time in history, is threatened by manmade changes to planetary systems – global warming, ozone depletion, desertification, toxic wastes, and acid deposition. Destruction of the ecosystem – globally, regionally, and locally – is possible as a result of human actions. *Economic growth*: Growth remains essential because the demographic explosion has not yet run its course. But growth must be redirected to be less taxing on nonrenewable resources and to avoid overloading disposal sinks. *Poverty and hunger*: These old curses of humanity have become endemic in some regions despite major development efforts. World-wide they afflict more people than ever before. Remedial action on all three fronts is possible. Sustainable development is the response to this multi-faceted challenge. It must become the overarching policy goal in order to keep the planet hospitable for a rapidly growing human population.

Having built this broad conceptual foundation the report is more cautious in its policy prescriptions. This is hardly surprising, given the delicate balancing act that was needed to reach consensus among commission members from all continents and countries with vastly different levels of development. Experienced policy makers served on the commission. Mrs. Brundtland, then Prime Minister of Norway, chaired the commission. William Ruckelshaus, twice administrator of the Environmental Protection Agency, was the U.S. member. Such people are not revolutionaries. They tend to recommend incremental change.

The Commission's central policy prescription is the integration of economic and environmental policies. "Environmental protection and sustainable development must be an integral part of the mandates of all agencies of governments, of international organizations, and of major private-sector institutions.   Ecological dimensions of policy [must] be considered at the same time as the economic, trade, energy, agricultural, industrial, and other dimensions – on the same agendas and in the same national and international institutions." (United Nations WCED 1987, pp. 312–313). Sustainable development, therefore, is not limited to "green" policies. It is much broader. It calls for a fundamental re-assessment, in theory and practice, of the interdependence between economy and ecology – going back to the common root both terms have in the Greek word "ecos," meaning household.

With this general principle in mind, five priorities are identified: (1) agreements designed to improve management of the global commons, the oceans, Antarctica, and outer space; (2) resolving conflicts among nations over scarce resources; (3) programs for and by developing nations to combat desertification, deforestation, environmental degradation and poverty; (4) programs by industrialized nations aimed at reducing threats from toxic chemicals, hazardous wastes and acid deposition; and (5) new international regimes to control global warming, ozone destruction, nuclear war, and international trade.

Our Common Future had an immediate worldwide impact. As an agenda-setting effort it was successful. Almost immediately a heated debate on the pros and cons of sustainable development began which continues to this date. A vast literature and innumerable conferences devoted to sustainable development attest to the fact that many people acknowledge the relevance and importance of sustainable development and its policy agenda. An unexpectedly broad coalition emerged. Not surprisingly, the idea of sustainable development appeals to environmentalists who see an opportunity to advance from newcomer to main player in the policy world. But the idea is also embraced by a growing number of business leaders. In 1990 the World Business Council for Sustainable Development was created. Two years later the Council published *Changing Course*, presenting sustainable development as a business opportunity rather than a threat (Schmidheiny 1992). Public agencies at all levels, international, national, regional, and local, have convened task forces to define sustainable development for their constituents.[1] Both industrialized and developing countries find common ground in sustainable development. While they differ sharply on the specifics of implementation, they agree with the analysis of underlying causes and the general direction of proposed action. This is an important asset for policy development and stands in stark contrast to the dogmatic confrontation, on concepts and theory, between East and West during the Cold War.

In 1992, the world summit on Environment and Development (UNCED) was held in Rio de Janeiro. This Third World meeting on the environment was entirely dedicated to debate and negotiation of sustainable development instruments, on climate change, biodiversity, and protection of rainforests. No country made binding commitments except to a timetable for continued negotiations. A massive conference document was also adopted, *Agenda 21,* that spells out sustainable policies and management practices for a wide range of human activities. In 40 detailed chapters all major policy arenas are reviewed and sustainable management and policy trends are identified. As a single illustration I cite the example of sustainable management of freshwater resources (Chapter 18). With personal experience in drafting policy documents that must survive the review process in international organizations, I expected a bland statement of general principles. Instead, I found specific information on such concepts as basin-wide management, integrated water resources development, water resources assessment, and others. The specialist will not find much new information here. But the intended audience is different. Water users and suppliers, from government, industry, and nongovernmental organizations, find guidance on what sustainable development means for their work. In particular local sustainable development projects benefit from information of this kind. Innumerable projects in river basins, irrigation districts and water utilities are underway all over the world searching for new ways of moving water management from an engineering to a user-driven task. *Agenda 21* gives stakeholders a good departure point. If the case of water management is applicable to other areas, the document has been successful in moving the principles of sustainable development into specific arenas of policy development and management. Obviously, information will need to be updated periodically in order to maintain its utility.

At the same time, ten years of debate have not elevated sustainable development to center stage in policy development. It certainly has not yet become a household word. Perhaps less so in rich than in poor countries. In developing countries, the urgency of some locally experienced sustainable development issues is great. For example, once the link between environmental destruction and poverty is recognized, incentives for reform become stronger and are often advocated under the banner of sustainable development.

In 1997, a review of progress since the original Rio Conference – "Rio Plus Five" showed little progress. Some dubbed the meeting "Rio Minus Five" to express the sense that momentum had been lost and that rich and poor countries alike were moving backwards.

Nowhere, so far, has sustainable development played well in electoral campaigns. Two small European countries, Norway and Holland, have approved "green" taxes. But in most countries voters do not yet see sustainable development as an urgent priority. Few political leaders tell them they should. Three are the exception. They identified their political future with the new policy agenda. Mikhail Gorbachev recognized, early during his tenure as leader of the Soviet Union, that the agenda of the Cold War needed to be replaced by the agenda of environmental and economic globalization (Gorbachev 1996, Gorbatschow *et al.* 1997). Albert Gore wrote *Earth in the Balance* when he served in the U.S. Senate (Gore 1992). Joschka Fischer, the intellectual leader of the German Green Party, wrote on globalization and the need for a new social contract in 1998 (Fischer 1998). Gorbachev paid a high price for his intellectual insight into changing world priorities. The Russian people became increasingly dissatisfied with his focus on global issues and felt that his priorities were wrong. It remains to be seen how Mr Gore will play the sustainability card, when he runs for President in 2000. The book by Mr Fischer is timed to support his candidacy for a ministerial appointment in the next German government. But Fischer produced more than a political manifesto. He carefully analyzes the connections between global environmental and economic changes. The author then proceeds to outline needed reforms in European labor and welfare policies to cope with these global challenges. All in all, a decade of agenda-setting debate has yielded a rich harvest of ideas, scholarship, popular writing, and public debate. On the other hand, critics complain that the concept of sustainable development remains excessively broad and provides little guidance for moving from idea to action.

Policy formulation

The list of new global challenges in need of a new policy response is long. I offer a simple and highly selective overview in Table 10.1. One can debate on how far to cast the net. Some would argue that global change as the challenge and sustainable development as the response are focused on

Table 10.1. *Examples of policy response to global change*

	Challenge	Response
Physical systems		
Atmosphere	Climate change	Climate Change Convention, Kyoto Protocol
	Ozone depletion	Vienna Convention. Montreal and London Protocols
Oceans	Resource depletion, pollution	Law of the Sea
Land	Desertification	UN Convention on Desertification
	Destruction of rainforests	Tropical Forestry Action Plan, Forest Principles
	Loss of biodiversity	Biodiversity Convention, Convention on International Trade in Endangered Species
Social systems		
	Population growth	Birth control programs. China, International organizations
	Poverty	Economic growth, welfare
	Urbanization/Megacities	
	Migration	
	Hunger, malnutrition	Relief programs
Economic systems		
	Trade	World Trade Organization/ GATT
	Labor	
	Property rights	World Intellectual Property and Copyright Treaty
	Over consumption	

man-made changes to physical systems. The Brundtland Commission, as summarized in the previous section, makes a strong case for taking a broader view. Direct linkages and powerful feedback loops exist among physical, social, and economic systems, but they are difficult to translate into policy. It is the very interdependence among systems that makes the policy response difficult. Other difficulties arise from the variety of political actors and the powers they wield. Only some issues can be addressed by the international community. Others are the legitimate domain of individual countries or their political subdivisions. Table 10.1 highlights international agreements. The titles of various policy measures point to the limited reach of the international policy system, where action is dependent on consensus among all parties and punishment of offenders is rare. "Principles" are declarations of intent. "Conventions" begin to specify an action agenda but remain broad. "Protocols" move on to binding commitments. The principles–convention–protocol sequence allows negotiations to move from declarations of intent to action that binds signatory nations. It often is a long and drawn-out process, and can be bypassed by any nation not willing or able to ratify the agreement (Susskind 1994). To illustrate international policy development I shall compare policy development against ozone depletion and global warming.

In 1995, the Nobel Prize in chemistry was awarded to three scholars whose laboratory work first established the link between the industrial uses of chlorofluorocarbons (CFCs) and the destruction of the atmospheric ozone shield, primarily over a wide area of the southern hemisphere. This manmade change to nature increases the amount of ultraviolet radiation reaching the surface of the earth. This can increase the number of skin cancers and may damage crops, forests, and ocean plankton. Some preliminary evidence also suggests increased risks to the health of livestock and humans.

The scientific work that was recognized by the 1995 Nobel Prize had been conducted in the early 1980s. Only a few years later satellite measurements confirmed the theory but also added the disturbing finding that the destruction of the ozone shield was progressing much faster than theoretical calculations had predicted. This resulted in fast action by the international community. By 1985 a general agreement was signed in Vienna. It identified the threat, reviewed the scientific evidence, and announced the intent to take international action. This was followed in 1987 by the Montreal Protocol under which CFCs were banned. A timetable for phasing out the offending chemical was part of the agreement.

Four years later, an additional protocol was signed in London offering financial and technical assistance to countries such as India, who had only recently begun to produce CFCs. The agreements worked. From 1994 to 1997, ozone depletion at midlatitudes in the northern hemisphere in winter and spring, for example, averaged 5.4 percent, almost a third less than the depletion that experts had projected several years ago.

The rapid chain of events in regulating CFCs suggests an easy path to the resolution of global environmental problems: Science detects a new problem that otherwise would have remained the unexplained cause of serious damage to various biological species. Armed with solid scientific information, rational decision makers lose no time in committing their nations to decisive action. They also help developing nations with the phase out of the harmful product. Industry already has developed a new product, which will pretty much do what CFCs did without harming the ozone shield.

The Montreal Protocol has been hailed as a model for global policy development. It is more likely, however, that this international agreement banning the use of a single industrial substance is the exception. It was an easy case. The science was not disputed. The threat was clear. Fewer than ten multinational corporations produced CFCs. The same companies had developed a substitute product that could be marketed without delay. Their income stream, therefore, was not threatened. Assistance to developing countries was small and not a major burden for the donor countries.

Since 1920, more than 150 international environmental agreements have been negotiated (Choucri 1993, p. xiv). Most have been long in the making and many of them are limited to easy actions on the part of the signatories – exchange of information and consultation among nations. Key countries have not ratified some of the more ambitious agreements, such as the Law of the Sea. No agreement contains the elaborate implementation and policing instruments that have been incorporated into the Nuclear Test Ban treaty. To this date no comprehensive regime for dealing with mankind's "global commons" – air, water and land – has been drafted. In the absence of world government such an attempt would be far beyond the capacity of existing international organizations. The contrast between the control of ozone depletion and greenhouse emissions – just two examples of new global challenges – illustrates the difficulty of reaching meaningful and timely international agreements when large economic interests are at stake.

The scientific base for linking man-made emissions from burning of fossil fuels to an increase in the natural greenhouse effect, and eventually to climate change, has been built over the course of an entire century. Exact measurements of changes in the atmospheric concentration of carbon dioxide, the main greenhouse gas, exist since 1957. From this time on, the numbers show how economic activities increase the amount of carbon dioxide and other greenhouse gases in the atmosphere. This point is not disputed. But debate and controversy, often in shrill tones and without respect for the facts, continue about the significance of this finding and the timing of possible consequences. An international assessment process, sponsored by the United Nations and conducted by the International Panel on Climate Change, has been underway for more than a decade. Stephen Schneider's chapter in this volume recounts the intense controversy surrounding the issue.

Debate and negotiations span about the same time period as in the case of ozone depletion. But meaningful action has not been taken. One of the earliest and most influential policy workshops on climate change concluded in 1985: "The understanding of the greenhouse question is sufficiently developed that scientists and policy-makers should now begin an active collaboration to explore the effectiveness of alternative policies and adjustments" (Villach Conference 1985). And, as early as 1988 Prime Minister Margaret Thatcher of Britain, in a speech to the Royal Society, correctly assessed the difficulty of the challenge:

> For generations we have assumed that the effects of mankind would leave the fundamental equilibrium of the world's system and atmospheric state unchanged. But . . . we have unwittingly begun a massive experiment with the system of the planet itself.

And she went on to endorse international controls of greenhouse emissions:

> even though this kind of action may cost a lot, I believe it to be money well and necessarily spent because the health of the economy and the health of the environment are totally dependent upon each other.
> (MARGARET THATCHER, September 27 1988. Speech to the Royal Society Dinner)

Ten years later the Kyoto agreement was reached in December, 1997. The intensity of the struggle is best illustrated by contrasting two statements that were presented to the delegates. The view of science, based on the 1995 assessment report by the International Panel on Climate Change, was summarized by the National Center for Atmospheric Research. The most important statement was that the human footprint was now discernible in measured climate events. Three markers of climate change were cited. Rapid melting of glaciers has been observed world-wide (except in Norway and Sweden, probably due to a larger snow pack). The last decade has been the warmest on record. Droughts and floods are becoming more severe.

Western Fuel Associates presented the opposing views of parts of the business community. These points were made: A small increase in warming will occur eventually. This should be welcomed because the world finds itself between two ice ages. Increased concentrations of carbon dioxide in the atmosphere will benefit agricultural production. The proposed climate treaty is based on computer simulations, not scientific observations. Mandatory greenhouse gas emissions will result in a major economic burden for the United States.

The main points of the Kyoto agreement are as follows: 38 industrial nations will reduce their greenhouse emissions by 5.2 percent by 2012. Higher rates apply to the European Union (8 percent), United States (7 percent), and Japan, Russia and Canada (6 percent). For the time being, developing countries are exempt from emission reductions. Industrial nations who finance emission reductions in developing countries will receive credit against their own emissions. As of this writing, the United States Senate is on record that it will not ratify a climate change treaty without emission reductions by all nations, including developing countries.

Even if this obstacle were removed, it seems unlikely that the international community is anywhere close to solving the greenhouse issue. Much higher reductions than those written into the Kyoto agreement are needed to stabilize the climate system. Uncertainty over climate change and its impacts continues to offer an easy reason for delaying action. The international community committed itself to prudent precautionary action on the basis of credible but not necessarily complete scientific evidence at the 1972 environment summit in Stockholm. Yet this principle of international relations has proven difficult to translate into binding agreements in other cases, such as acid deposition and loss of biodiversity.

Climate change is more complex by far. Virtually all of us are asked to change our lifestyles by giving up current amenities for possible future benefits. The cost-benefit calculation that all of us make instinctively is far from supporting a policy consensus in favor of higher gasoline prices, a carbon tax, or the electric automobile. Economic interests tied to the fossil fuels industries may be the most vocal opponents, but they have plenty of company. Years ago Tom Schelling (1983) put it this way (and the situation has not changed fundamentally since that time):

> In the current state of affairs the likelihood is negligible that the three great possessors of the world's known coal reserves – the Soviet Union, the Peoples Republic of China, and the United States of America – will consort on an equitable and durable program for restricting the use of fossil fuels through the coming century and successfully negotiate it with the world's producers of petroleum and with the fuel-importing countries, developed and developing.
>
> (SCHELLING 1983)

Schelling concluded that a workable climate agreement is out of reach: the entire industrial and transportation system would need to be restructured. And even if this were done, an agreement would be difficult to police. The possibilities for cheating would be endless.

Technological alternatives to regulatory control of emissions exist, such as development of additional carbon sinks (forests) and carbon sequestration in the ocean or underground. The transition to energy from non-fossil fuels would re-stabilize the climate system. Yet policy development along these lines, with the exception of modest research and development funding, remains absent. Restructuring of the tax system by reducing income taxes and imposing taxes on the carbon content of industrial products has been proposed. With the exception of small European countries these reforms have not found sufficient political support. For the time being, climate change policy remains marginal and ineffective. The same is true in other policy arenas, such as the protection of tropical rainforests, the oceans and biodiversity.

Three conditions for successful policy formulation have not yet been met: First, perception of risks must be widespread. So far, the risks of inaction are not yet immediate or high enough to allow a political majority to emerge that will support control of emissions. Second, leadership for change is spotty and cautious. Being in favor of sustainable development is not the same as endorsing a carbon tax. Third, the proposed policy instruments seem complex and without direct appeal to the majority of voters.

A new policy paradigm?

New policies take time to be born. Revisit for a moment a previous challenge, the first industrial revolution, and the response to social dislocations brought by it, the welfare state. The medieval welfare system was based on family, community, and church. This system collapsed. It could not provide for retirement benefits, unemployment assistance, and accident insurance of the new industrial workforce. It took about 50 years into the industrial revolution before England passed the first social legislation. Germany followed 50 years later. The United States acted only during the great depression of the 1930s. Over time, all industrialized countries accepted the need for public support of a safety net against sickness, old age, disability and loss of the breadwinner.

A second policy role also emerged when the state began to curb abuses of industrial power. Antitrust, monopoly, consumer protection and safety legislation was passed, again over many decades. These two bodies of law, protection for the industrial worker and family, and regulation of industry, make up the welfare state. Powerful groups advocated, often forced, the development of these policies. Trade unions and new political parties dedicated themselves to this task. Temporary coalitions, like the progressive movement, changed the face of American political institutions around the turn of the last century. Labor unrest, strikes, at times political revolutions, marked the process. The causes people fought for were easily understood: A decent wage, protection against unforeseen events during the working life, safety in the workplace and fair deals offered by industry in selling its products and services. By the middle of this century the welfare state of the key industrial nations was in place. It had taken over a century to reach this point, to overcome many false starts, and to learn from experience. The Soviet experiment to join welfare and industrial power – the planned rather than incremental search for solutions – collapsed after 70 years.

This reminds us how slow, messy and costly policy development tends to be. We should not expect this to be different in the case of sustainable development. Yet the process may also be different, and even more difficult, than in the past. I see three reasons for this to be the case.

First, the goals of sustainable development are less immediate and direct. It will be more difficult, therefore, to gain the support of the majority. So far, sustainable development is more a concern of the educated elite than of the masses.

Second, many of the problems of the twenty-first century cannot be detected by the naked eye. We need science to define, measure and assess the depletion of the ozone shield, or the progress of climate change. This makes the policy process more difficult to understand and moves it away from the majority of people.

Third, often we do not know whether environmental degradation will proceed gradually in a linear fashion or in an abrupt manner once a threshold is reached. The ozone case surprised scientists with its unexpectedly fast progression. Policy making proceeds by trial and error; it is social learning. We rarely get it right the first time around. It took several generations to build the welfare state. Will we have enough time to build the sustainable state?

These observations seem excessively cautious about the chances of sustainable development as the new policy paradigm. They are meant to be this way because the tasks ahead are huge, poorly understood, and urgent. I see no evidence that sustainable development is emerging as the dominant policy paradigm for international and national actions. China's population policy could be cited as the single most significant counter example. The one-child-per-family policy was indeed successful. It will make India, not China, the most populous state early in the twenty-first century. But the policy was enforced with draconian measures. Is the solution worth the cost? Could it be that sustainable development is within the grasp of an authoritarian state only? These are disturbing questions.

Revolution at the grassroots

There is, however, encouraging evidence that sustainable development as a new policy paradigm is taking root at the bottom of the political system. To illustrate challenge and response in the local and regional context I offer two quotes from the best book on policy and sustainability that I have come across.

Charles Wilkinson is a specialist of natural resource law and policy in the American West. He describes the laws of the West as products of their time, made for a specific purpose to serve a widely recognized need. Over time, they have lost their utility. Some have become counter-productive. A society and its political system will perish unless obsolete laws are abandoned and replaced by more appropriate provisions to protect the common good. In five case studies on mining, grazing, logging, fishing, and water use Wilkinson tells the story of how laws and policies came into being, served their purpose, and now stand in the way of sustainable development in the American West.

> A consensus exists that western resources generally ought to be developed but that development ought to be balanced and prudent, with precautions taken to ensure sustainability, to protect health, to recognize environmental values, to fulfil community values, and to

provide a fair return to the public. These principles have broad acceptance, but development in the West does not proceed in accordance with them. Rather, westwide, natural resource policy is dominated by the lords of yesterday, a battery of nineteenth-century laws, policies and ideas that arose under wholly different social and economic conditions but that remain in effect due to inertia, powerful lobbying forces, and lack of public awareness.

(WILKINSON 1992, p. 17)

What will it take to get rid of the lords of yesterday and replace them with rules and practices that are more sustainable? Wilkinson's answer will surprise many.

It seems to me inevitable that westerners increasingly will turn to various forms of planning . . .: the process of a community coming together; identifying problems; setting goals – a vision – for a time period such as twenty or forty years; adopting a program to fulfill those goals; and modifying the program as conditions change.

(WILKINSON 1992, p. 300)

I am comfortable with this bottom up approach to policy development. Having observed the dedication of local groups working on their own future in Brazil, Mexico and Texas, I can attest to the political power and democratic spirit of stakeholder planning. This is the way to change minds and train citizens in sustainable development. However, the process takes time before it reaches critical mass. The growth of non-governmental organizations, a world-wide phenomenon, is an important institutional innovation that facilitates this grassroots approach (Salamon and Anheier 1994). Collective action and civic action, as Robert Putnam has shown in his study on Italy, are central to successful democracy (Putnam 1993). Policy development for sustainable development may get its strongest push from community action. Over time, this can build the constituency needed to tackle sustainability issues on the scale of nations and the international community.

The grassroots approach to sustainable development also entails a powerful partnership with science. Because the issues at hand tend to have large technical components, scientists are called upon to diagnose problems and define the range of policy options. Stakeholders choose among solutions and make policy recommendations. Solutions are based on intensive dialogue between experts and stakeholders. Civic science joins expert knowledge to sustainable development projects and policies (Schmandt 1998). Holistic scientific inquiry serving community goals can pave the road to sustainable development.

REFERENCES

Choucri, N. 1993. *Global Accord: Environmental Challenges and International Responses.* MIT Press, Cambridge, Massachusetts.

Fischer, J. 1998. *Fuer einen Neuen Gesellschaftsvertag Eine politische Antwort auf die globale Revolution.* Kipenheuer & Witsch, Koeln, Germany.

Gorbachev, M. 1996. *Memoirs.* Doubleday, New York.

Gorbatschow, M., Sagladin, V. and Tschernjajew, A. 1997. *Das Neue Denken: Politik im Zeitalter der Globalisierung.* Goldman Verlag, Munich, Germany.

Gore, A. 1992. *Earth in the Balance: Ecology and the Human Spirit.* Plume/Penguin, New York.

Putnam, R. D. 1993. *Making Democracy Work: Civic Traditions in Modern Italy.* Princeton University Press, Princeton, New Jersey.

Salamon, L. M. and Anheier, H. K. 1994. *The Emerging Sector: The Nonprofit Sector in Comparative Perspective.* The Johns Hopkins University, Institute for Policy Studies, Baltimore, Maryland.

Schelling, T. C. 1983. Climate change: implications for welfare policy. In National Research Council, *Changing Climate.* National Academy Press, Washington, DC.

Schmandt, J. 1998. Civic Science. *Science Communication* 20(1):62-69.

Schmidheiny, S. with the Business Council for Sustainable Development. 1992. *Changing Course: A Global Business Perspective on Development and the Environment.* MIT Press, Cambridge, Massachusetts.

Susskind, L. E. 1994. *Environmental Diplomacy: Negotiating More Effective Global Agreements.* Oxford University Press, New York.

United Nations World Commission on Environment and Development. 1987. *Our Common Future.* Oxford University Press, Oxford and New York.

Villach Conference. 1985. From the Concluding Statement of the Villach Conference; World Meterological Organization–United Nations Environment Programme, 1988; *Developing Policies for Responding to Climatic Change: A Summary of the Discussions and Recommendations of the Workshops Held in Villach and Bellagio.* WCIP-1, WMO/TD-No. 225. World Meteorological Organization, Geneva.

Wilkinson, C. F. 1992. *Crossing the New Meridian: Land, Water and the Future of the West.* Island Press, Washington, DC.

ENDNOTE

1. See, for example, *Sustainable Development: OECD Policy Approaches for the 21st Century.* Paris: Organization for Economic Co-Operation and Development, 1997. *Sustainable America: A new consensus for prosperity, opportunity, and a healthy environment for the future.* Washington, DC: President's Council on Sustainable Development, 1996.

Index